D0854565

Global Warming, River Flows and Water Resources

A volume in the *Water Science Series*

Published on behalf of the Institute of Hydrology

GLOBAL WARMING, RIVER FLOWS AND WATER RESOURCES

Nigel Arnell
University of Southampton

JOHN WILEY & SONS
Chichester • New York • Brisbane • Toronto • Singapore

Copyright © 1996 by the Institute of Hydrology
Crowmarsh Gifford
Wallingford, Oxon
OX10 8EX, England

Published by John Wiley & Sons Ltd,
Baffins Lane, Chichester,
West Sussex PO19 1UD, England

National   01243 779777
International (+44) 1243 779777
e-mail (for orders and customer service enquiries): cs-books@wiley.co.uk
Visit our Home Page on http://www.wiley.co.uk
or http://www.wiley.com

Other Wiley Editorial Offices

John Wiley & Sons, Inc., 605 Third Avenue,
New York, NY 10158-0012, USA

Jacaranda Wiley Ltd, 33 Park Road, Milton,
Queensland 4064, Australia

John Wiley & Sons (Canada) Ltd, 22 Worcester Road,
Rexdale, Ontario M9W 1L1, Canada

John Wiley & Sons (Asia) Pte Ltd, 2 Clementi Loop #02-01,
Jin Xing Distripark, Singapore 129809

British Library Cataloguing in Publication Data

A catalogue record for this book is available from the British Library

ISBN 0-471-96599-5

Produced from camera-ready copy supplied by the Institute of Hydrology
Figures redrawn by VIZIM
Printed and bound in Great Britain by Bookcraft (Bath) Ltd.
This book is printed on acid-free paper responsibly manufactured from
sustainable forestation, for which at least two trees are planted for each one
used for paper production.

1001342379

Contents

For Hilary: thanks for all the support

Preface

In May 1996, as the final touches were being put to this book, reservoirs in the south Pennines were holding less than 15% of their capacity: normally at this time of year they would be close to full. Newspapers and television bulletins talked of the chance of drought during the summer — and of the additional threat of global warming.

This book explores the possible consequences of global warming for river flows and water resources, concentrating on Britain. It is based around a state-of-the-art climate change scenario, under which water resources in the south and east of Britain are reduced in the future whilst resources in the north and west are increased. The variability in flow through the year is generally increased, with proportionately more occurring during winter and less during summer. Coupled with increased demand for water due to higher temperatures, water resources across much of Britain would be under greater stress in the future as a result of global warming. The book critically evaluates methodologies for estimating the impacts of global warming on hydrology and water resources, and also identifies the major uncertainties: strategies for adapting to global warming are also explored.

Dr Nigel Arnell is currently Reader in Physical Geography at the University of Southampton. Until 1994, he worked at the Institute of Hydrology, specialising in the effects of global warming on river flows, regional hydrology and flood estimation. He was a lead author for the Intergovernmental Panel on Climate Change (IPCC) Second Assessment report on the impacts of climate change, published in 1996, and is a member of the UK Department of the Environment Climate Change Impacts Review Group (CCIRG). He has advised the National Rivers Authority on the implications of global warming, and leads a European Commission-funded project investigating climate change and water resources in Europe.

CHAPTER 1

Global warming, river flows and water resources

THE CONTEXT

Global warming became one of the biggest scientific issues during the 1980s. It has continued to attract scientific attention and also political and public concern. The drought that affected much of Europe between 1988 and 1992 (and lasted even longer in southern Europe) was attributed in some quarters to the "greenhouse effect", as was the drought in Britain in the summer of 1995. Record flooding in many British rivers in 1994 and 1995, and in the Rhine basin, was also blamed by some on global warming. But is it yet possible to see any evidence of global warming, and what future effects might there be?

The broad aim of this book is to review the possible effects of global warming on river flows and water resources in Britain. It is derived from a case study of potential changes to river flows in British catchments, but a context for the British results is provided by an analysis of studies in other environments.

Global warming is the name given to the possible climatic effect of an increasing concentration of "greenhouse" gases, primarily carbon dioxide (CO_2), methane and nitrous oxide. In the most general terms, these gases are transparent to incoming short-wave radiation, but they block outgoing long-wave radiation, leading to an increase in both surface radiation and surface temperature. These increases in turn, it is hypothesised, lead to changes in climate. Concentrations of the greenhouse gases — so named because they act like the panes of glass in a greenhouse — have been increasing as a result of human activities such as the burning of fossil fuels, deforestation, the application of fertilisers and the growth of particular types of agricultural production. Water vapour is also a greenhouse gas: although its concentration in the atmosphere is not directly affected by human activity, it does increase with temperature, and therefore tends to reinforce the effect of the other greenhouse gases.

There are two principal reasons for attempting to estimate the impacts of future global warming on hydrological and water resources systems. The first is to provide scientific information for policy-makers and managers of systems exposed to climatic variability over the next few years or decades. The second is to provide scientific information to inform public debate and to constitute a basis for political decisions on the mitigation of global warming. Under the Framework Climate Change Convention, introduced at the United Nations Conference on Environment and Development (UNCED) in Rio de Janeiro in June 1992, strategies to reduce or limit global warming should be based on an assessment of the impacts of climate change.

Before going any further, it is important to clarify two terms. *Climate change* is the general term for any persistent change in climate, occurring over decades. *Global warming* is one particular type of climate change. Unfortunately the term *climate change* has sometimes been taken to be synonymous with *global warming*. This is slightly misleading, since a decline in temperature before an ice age would also be climate change. This book is concerned specifically with global warming, and any references to "climate change" made within it should be taken to mean change associated with an increasing concentration of greenhouse gases.

RESEARCH INTO GLOBAL WARMING AND RIVERS

The possibility that increased concentrations of carbon dioxide in the atmosphere could lead to higher temperatures was first raised by the Swedish scientist Arrhenius in the late 19th century, but it was not until the late 1970s that scientific concern about global warming really developed. This reflected not just an observed increase in temperature, but also theoretical and methodological developments which made it possible to estimate the effects of increasing concentrations of carbon dioxide, and other greenhouse gases, on climate.

The international scientific and political community was sufficiently concerned about the potential effects of future global warming, and the apparent evidence provided by increasing global temperatures (Figure 1.1), to establish the Intergovernmental Panel on Climate Change (IPCC). This was set up by the World Meteorological Organization and the United Nations Environment Programme in 1988, with the aim of assessing the possible chances, effects and responses to global warming. The IPCC works through groups of experts, and reports have been prepared on climate change processes (IPCC, 1990a; 1992), impacts (IPCC, 1990b, 1993) and possible responses (IPCC, 1991). The IPCC Second Assessment — covering science, impacts and policy — is to be published during 1996. The IPCC represents

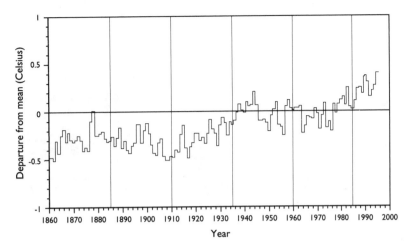

FIGURE 1.1   Variation in global temperature, 1860 to 1995. Data courtesy of the Hadley Centre for Climate Prediction and Research.

the scientific state of the art, and its results will be used by governments and international agencies to frame and support appropriate response policies.

The first comprehensive studies into the potential effects of climate change on water resources date back to 1977, when a report was commissioned by the United States National Research Council (National Research Council, 1977). The next significant publication was that of Nemec and Schaake (1982), which simulated the effects of changes in temperature and precipitation on river flows and reservoir operation in several example catchments. Since then, there has been a rapid expansion of research into climate change effects. Figure 1.2 shows the number of papers on climate change and its effects on hydrological regimes and water resources published in several key journals.

These studies have followed a range of approaches with different objectives. Most have been case studies into the effects on hydrological regimes in different catchments, while few have looked at the implications for water resources. Fewer still have expressed the possible impacts in financial or monetary terms, and only since the early 1990s have there been many studies looking at the effects of global warming on the aquatic environment.

The first general conclusion that can be drawn from these studies is that the effects of climate change on river flow regimes depend not only on climate change, but also on the current climate of the catchment and on its physiographic properties. A given climate change can have a completely different effect in different catchments. A second general conclusion is that different scenarios can produce very different impacts in any given catchment, and that the hydrological system tends to exaggerate the differences between scenarios. These two conclusions will be expanded upon in later chapters.

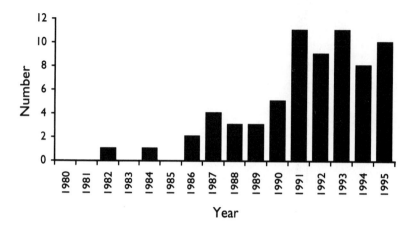

FIGURE 1.2   Number of papers on the effects of future climate change on hydrological regimes and water resources, 1980 to 1995. The following journals were included: *Water Resources Research, Journal of Hydrology, Hydrological Sciences Journal, Water Resources Bulletin* and *Climatic Change.*

STRUCTURE OF THE BOOK

This book is built upon a case study into the effects of global warming on river flows in British catchments. It is not intended to review the processes behind global warming in any detail, nor to discuss measures for mitigating its effects. Chapter 2 introduces the science of global warming and outlines the range of potential effects on hydrological processes and on the components of the water balance. It also compares climate change with climatic variability and considers the difficult problem of detecting a global warming signal in hydrological data.

Chapter 3 introduces the methodology behind a climate change impact assessment study, drawing heavily on the work of the IPCC (Carter *et al.*, 1994). It explores suitable methods of creating future climate change scenarios at space and time scales appropriate to hydrological studies: this is perhaps the single most important technical stage in an impact assessment, since the credibility of estimated future hydrological patterns is largely determined by the credibility of scenarios for changes in the controlling climate variables.

The British case study is introduced in Chapter 4, which summarises the catchments used and hydrological models applied. The case study uses climate change scenarios produced for the UK Department of the Environment Climate Change Impacts Review Group (CCIRG, 1996; Hulme, 1996), which has reviewed the potential effects of global warming across all sectors of the UK economy. The scenarios are based on output from the Hadley Centre transient climate change experiment (Murphy, 1995). The case study is a development, using a different scenario, of a study conducted for the UK Department of the Environment (Arnell and Reynard, 1993; 1996).

Chapter 5 considers the effect of global warming on river flow regimes and groundwater recharge, largely through the British case study but also by reviewing studies in other areas. It looks at changes in annual and monthly average runoff and annual average recharge by the 2050s, and at changes in the variability of flows through the year and the occurrence of high and low flow extremes. It considers the aspects of change and the catchment which influence response to global warming. Chapter 6 looks at the transient effects of a gradual change in climate over the next few decades, and compares an underlying trend with year-to-year and decade-to-decade variability. Effects of global warming on river and lake water quality are reviewed in Chapter 7, which relies heavily on a study undertaken by Jenkins *et al.* (1993).

The final chapter attempts to draw some inferences for water resources and their management in Britain from the potential changes in water quantity and quality. It considers changes in the demand for water, and then examines different water uses — including water supply, power generation, irrigation, navigation, recreation and instream use by aquatic ecosystems — before exploring potential effects on water as a hazard (flood, drought and ill-health). The chapter explores the implications for water management, discusses the effects of adaptation to change (to which the water industry is well accustomed) on the impacts of global warming, and finishes with some suggestions of areas for future research.

Chapter 2

Climate change and hydrological processes

This chapter defines the greenhouse effect and global warming, examines their potential implications for hydrological processes and the hydrological cycle, and finally considers whether a global warming signal can be detected in hydrological data. First, however, it is important to provide some context for the discussion of future climate change by looking at past variations and exploring the differences between climate variability and climate change.

CLIMATIC VARIABILITY AND CLIMATE CHANGE

Weather describes the day-to-day state of the atmosphere; climate is defined as average weather. Climate, however, is not constant, but varies over many time scales (Figure 2.1).

The earth's climate has varied since the beginning of time. Over the last two million years the dominant pattern is the Pleistocene glacial/interglacial cycle, with a period of around 100 000 years (Imbrie and Imbrie, 1979). Global average temperatures varied by 5 — 7°C between glacials and interglacials, and average temperature in some middle and high latitude regions in the Northern Hemisphere varied by 10 — 15°C. The reason for the glacial-interglacial cycle is believed to be small changes in summer radiation receipts, caused by periodic variations in the earth's orbit (the Milankovitch cycle: IPCC, 1990a). The earth emerged from the latest glacial period between 10 000 and 15 000 years ago: during the period since then (the Holocene) there have been a number of significant events.

Around 10 500 years ago the Holocene warming was abruptly reversed within about a century. The cold spell persisted for approximately 500 years and ended as abruptly as it began (Dansgaard *et al.*, 1989). This event — the Younger Dryas event — cannot be attributed to orbital changes, and the most likely

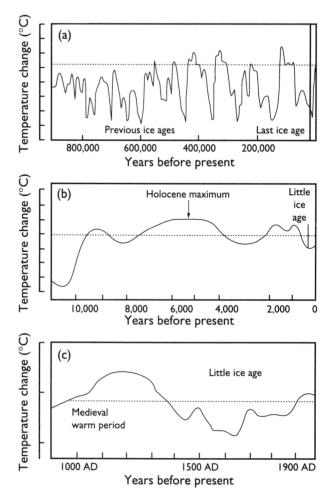

FIGURE 2.1   Variation in global temperature over different time scales (IPCC, 1990a)

theory is that it was associated with the sudden influx into the North Atlantic of dense, cold freshwater following large-scale melting of the Laurentide Ice Sheet in North America and its drainage through the St Lawrence River, rather than south through the Mississippi basin (Broecker *et al.*, 1985). This would have altered ocean circulation patterns in the North Atlantic, and possibly global oceanic circulation too, with consequent effects on climate.

The warmest period since the end of the last glacial period was between 5000 and 6000 years ago — the Holocene "climate optimum" — when temperatures in mid latitudes were 1-2 degrees warmer than at present (IPCC, 1990a). Since this high value, temperatures have continued to fluctuate at different time scales and three events are particularly important: the Medieval warm epoch, the Little Ice Age and the period of warming during the 20th century (considered in more detail later in the chapter). In Europe, the Medieval warm epoch lasted from around AD 1000 to 1300 (Lamb, 1982). Mean temperature in Britain during the 13th century could have been 0.4 — 0.8 degrees warmer than the preceding and following periods. This period was one of agricultural expansion, including the northward spread of viticulture (Lamb, 1982). The Little Ice Age reached a peak between 1650 and 1750 (in Europe at least) and saw glacial advances, harvest failures and increased flooding in many parts of Europe (Hulme, 1994a). These fluctuations, superimposed on the glacial-interglacial cycle, show considerable geographical variability, and are not necessarily synchronous. There is still debate about the causes of these, and the many shorter-lived fluctuations, but possible contributing factors include slight variations in the output of solar radiation, inherent thresholds within the climate system (such as the ice melt threshold, which was probably responsible for the Younger Dryas event) and irregular volcanic eruptions.

The Little Ice Age is at the very beginning of systematic climate observations. Manley (1974) began to compile a record of temperature in central England, starting at a monthly resolution in 1659. Annual, winter and summer temperatures are shown in Figure 2.2. A similar record of precipitation for the period since 1766, averaged over a number of recording sites, has been constructed by Wigley et al. (1984) (Figure 2.3). Despite some problems with the data (Hulme, 1994a), both these sets of graphs show the variability in British climate over the past few hundred years. For example, when averaged over a few years, temperature varies within ±0.5° of the long-term mean, and multi-year average precipitation can vary by up to 10% of the long-term mean.

It is rather more difficult to produce consistent time series of hydrological properties. Very long time series of lake levels can be recreated from sedimentological, morphological and palynological evidence (Street-Perrot and Roberts, 1994), but it is considerably harder to estimate continuous series of river flows from indirect evidence. The studies that have reconstructed records of river behaviour back through geological time have tended to identify periods of high flows, rather than continuous series, either from sedimentological evidence or from changes in river form (Knox, 1995; Starkel, 1995). During the historical period documentary evidence can be found of hydrological extremes, again focusing largely on notable flood events, and it is possible to construct flood chronologies.

The earliest surviving hydrological record is from the Nile at Cairo, dating back to 641 AD. Unfortunately there are many gaps between the 16th and 18th centuries, and although it is possible to identify some anomalous periods coinciding with climatic anomalies in Europe (Hassan, 1981), there are few data from the Little Ice Age. The length of the Nile record is

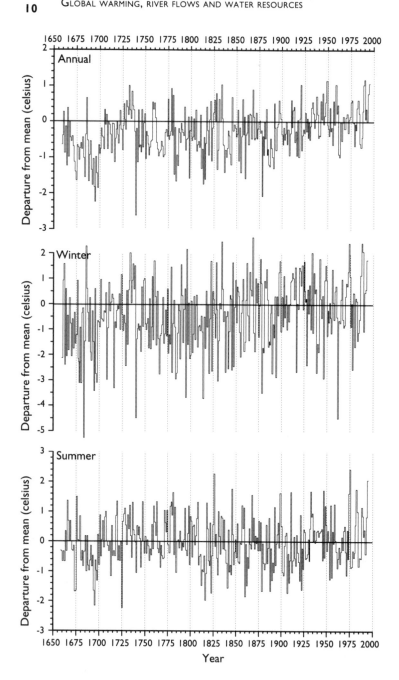

FIGURE 2.2   Central England temperature from 1659 to 1995, expressed as a departure from 1961 — 1990 average (Manley, 1974). Data courtesy of the Climatic Research Unit, University of East Anglia.

exceptional: no other consistent records extend nearly as far back. During the early 19th century river levels and flows began to be recorded systematically on many large European rivers. Flows were first measured in Britain on the River Colne in 1834 (Marsh, 1996), but the longest continuous flow measurements started in 1879 on the Lea in Essex. The major problem with recorded flow data is that conditions in the

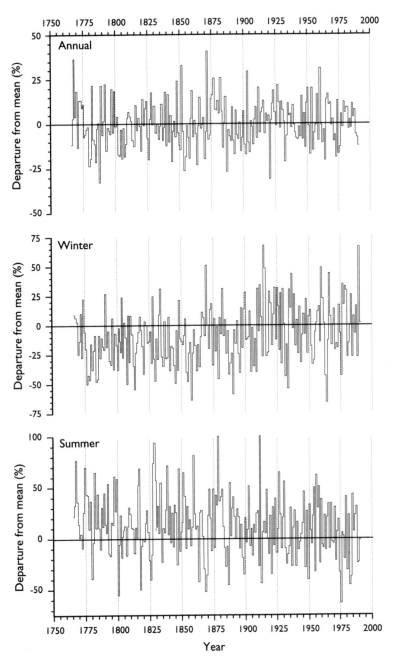

FIGURE 2.3　England and Wales rainfall from 1766 to 1992 expressed as percentage departure from 1961 — 1990 averages (Wigley et al., 1984). Data courtesy of the Climatic Research Unit, University of East Anglia.

catchment may have changed significantly during the period of record. Not only has catchment vegetation often changed, but reservoirs might have altered the volume and timing of flow, and channel works might have significantly affected the flood hydrograph. It is therefore very difficult to produce consistent long hydrological time series stretching back more than about 50 years.

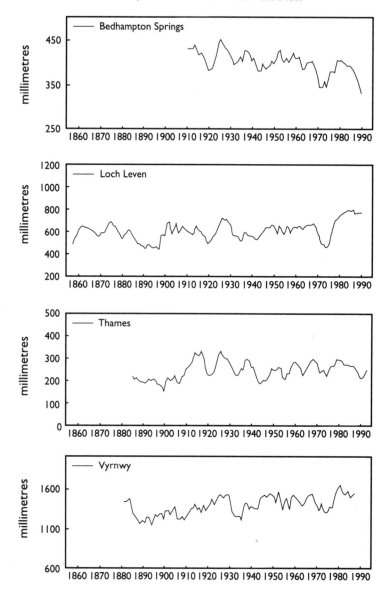

FIGURE 2.4   Variations over time in river flows in four British catchments with long records (Marsh, 1996)

It is, however, possible to reconstruct "natural" river flow records over the past few hundred years using various techniques — for example correlation with precipitation data or tree-rings (Jones *et al.*, 1984; Young, 1994) — and it is frequently possible to derive estimates of past river flows from records of lake levels or spring flows. Figure 2.4 shows five-year running means of annual river totals at four sites in Britain (Marsh, 1996). The Loch Leven record from Scotland represents inflows to the loch, reconstructed from information on levels and sluice records (Sargent and Ledger, 1992). The Bedhampton

Spring record is derived from observations taken over many years at a spring on chalk in Hampshire. The Thames record is based on data recorded at Teddington Weir and Kingston (on the western edge of London) since 1882, whilst the Vyrnwy record from Wales is based on reservoir levels. All four series clearly show fluctuations over time, often but not always synchronous.

This section has introduced the concept of climatic variability, particularly the superimposition of shorter-term fluctuations onto the 100 000 year glacial-interglacial cycle. Clearly, it is not appropriate to conceive of climatic characteristics, or hydrological properties, as constant over long periods: the graphs in this section have revealed periods in the past with sustained conditions above, or below, the long-term average, followed by return to "average" conditions. Figure 2.2 in particular also hints at the greenhouse effect, which may be a *change* superimposed on this considerable variability.

THE GREENHOUSE EFFECT

The physics of the greenhouse effect: an introduction

The earth's climate system is driven by short-wave energy received from the sun. A portion of this incoming radiation is reflected from clouds and at the earth's surface: the balance heats the ocean and land surface. Warm surfaces re-radiate energy at longer wavelengths. This long-wave radiation is absorbed by certain trace gases in the lower atmosphere, thus causing the air temperature to rise (Figure 2.5). A much more comprehensive description of the greenhouse effect can be found in the IPCC reports (IPCC, 1990a; 1992). Without this effect (known as the "greenhouse effect" because the panes of glass in a greenhouse act in a similar way to the trace gases in the atmosphere) the temperature of the earth would be 33°C cooler and life would be unsustainable.

The most important of the radiatively active trace gases are water vapour and carbon dioxide. These greenhouse gases have become more abundant in the atmosphere as a result of human activity, and new man-made greenhouse gases have been introduced (Table 2.1). The increase in CO_2 mainly results from increased combustion of fossil fuels and deforestation. Rates of emission depend on economic development and the efficiency of energy use, and can be estimated and modelled with reasonable accuracy. However, some of the sinks in the carbon cycle — principally vegetation, soils and the oceans — are very poorly understood, and there is considerable uncertainty in the rate of change of atmospheric carbon dioxide concentrations. Methane is produced by a variety of anaerobic

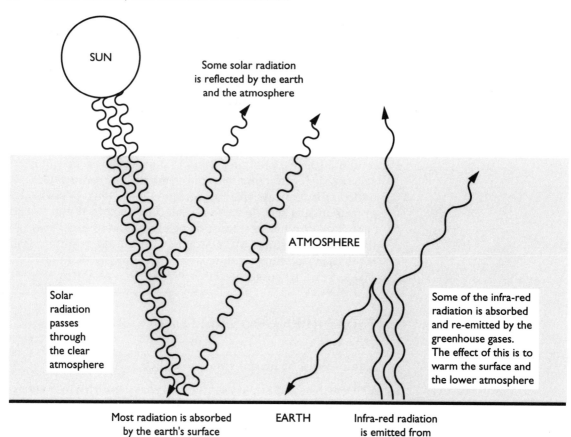

FIGURE 2.5   The greenhouse effect (IPCC, 1990a)

TABLE 2.1   Greenhouse gases in the atmosphere (IPCC, 1990a)

	Pre-industrial concentration	Present (1990) concentration	Annual rate of increase	Major sources
Carbon dioxide	280 ppmv	353 ppmv	0.5%	fossil fuels combustion, deforestation
Methane	0.8 ppmv	1.72 ppmv	0.6%*	rice production, cattle rearing, industry
Nitrous oxide	288 ppbv	310 ppbv	0.25%	internal combustion engine, agriculture
Tropospheric ozone	no data	no global estimates	no data	
CFC-11	0 pptv	255 pptv *	4%	aerosols, refrigeration
CFC-12	0 pptv	453 pptv *	4%	aerosols, refrigeration

* values as revised in IPCC (1992)

(oxygen deficient) processes: the major human-influenced sources are paddy rice and ruminants. Concentrations have also increased because of biomass burning, coal mining and the venting of natural gas. Combustion of fossil fuels may also lead to a reduction in the rate of certain zone is not emitted directly by human activities, but is formed in the atmosphere by photochemical transformations of gases such as carbon monoxide and nitrogen oxides which do have human origins. Chlorofluorocarbons (CFCs) are (molecule for molecule) very powerful greenhouse gases, but their enhanced greenhouse effect is largely offset by their effect in removing stratospheric ozone (IPCC, 1992).

Water vapour is the most important greenhouse gas. It is not affected directly by human activity, at the global scale, but higher temperatures caused by the increased concentrations of other greenhouse gases would cause water vapour concentrations to increase. This is an example of one of the positive feedbacks that characterise the atmospheric system.

Sulphate aerosols are a product of atmospheric pollution, and tend to reflect incoming solar radiation back out into space (Mitchell *et al.*, 1995a, b; Ericson *et al.* 1995). They therefore counteract the greenhouse effect to some extent, particularly in the Northern Hemisphere where most of the sulphate aerosols are produced, but their effect is highly dependent on their spatial distribution. Improvements in air quality would therefore enhance the greenhouse effect by reducing sulphate aerosol concentrations.

Modelling the regional effect of increasing greenhouse gas concentrations

Increases in greenhouse gas concentrations alter the energy budget in the lower atmosphere, resulting in higher temperatures, which in turn mean changes in global and regional climate patterns. However, there are many unknowns at each stage of the chain, making it difficult to predict the effects.

The concentrations of greenhouse gases in the atmosphere depend on their rate of emission by human activity, their interactions within the atmosphere, and the rate at which they are taken up by natural sinks. The estimated effect of a given concentration of greenhouse gases on temperature depends on the understanding and simulation of atmospheric and oceanic processes: the role of clouds in reflecting or absorbing radiation is particularly important. Regional effects of global warming can be simulated only by using three-dimensional physically-based representations of atmospheric and oceanic processes.

Global Circulation Models (GCMs) are based on the laws of physics, and simulate the dynamic behaviour of short-term weather patterns. GCMs simulate the large-scale characteristics of the present general circulation reasonably well, although all of them simulate temperature better than precipitation, and may be highly inaccurate for particular regions. They operate at a short time scale (down to 15 minutes) but a coarse spatial resolution. The UK Hadley Centre High Resolution GCM, for example, has a spatial resolution of 3.75 × 2.5°, or approximately 80 000 km^2 in Europe. This spatial scale is adequate to simulate large-scale climatic processes, but is too coarse to capture the local meso-scale forcings, which not only affect local and regional climates, but may also feed into large-scale circulation patterns. These meso-scale forcings include topography, the coastline (setting up thermal contrasts between warm sea and cold land), inland water and vegetation patterns (Giorgi and Mearns, 1991). Climate models simulate the effects of sub-grid-scale processes through parameterisations of these processes. These parameterisations should also incorporate the spatial variability in these processes, but unfortunately there are few data to help refine them.

Considerable research effort is being devoted to improving GCMs: ocean-atmosphere interactions are being improved (IPCC, 1992), cloud parameterisations are being updated (IPCC, 1992) and spatial resolution is being increased. Hydrological processes at the land surface have an important effect on the partitioning of energy between latent and sensible heat. They are the subject of major field experiments (Goutorbe et al., 1993; Andre et al., 1988; Sellers et al., 1992) and modelling studies (e.g. Laval and Polcher, 1993; Pitman, 1993; Vorosmarty et al., 1993; Stamm et al., 1994).

EFFECTS ON HYDROLOGICAL PROCESSES

Figure 2.6 summarises the impact of a change in greenhouse gas concentrations on the natural hydrological system, emphasising the relationships between all elements of the system (Arnell, 1994). The increase in greenhouse gas concentrations results in an increase in radiation at the surface, increasing temperature. Higher surface temperature produces a change in rainfall and evaporation, which together change river flows and ground-water recharge. Changes in temperature, radiation, rainfall soil moisture and CO_2 concentrations all affect catchment ecosystems and land use, which then affect the catchment water balance. Stream water quality is affected by temperature, catchment land use, rainfall, the volume of river flow and saline intrusion.

Although the greatest attention has been directed towards changes in the mean climate, the variability of climatic inputs to the hydrological system are also likely to change. A simple rise in the mean of an input — such as precipitation — would affect the mean value of the output — such as streamflow —

and could also affect the variance of the output because of the non-linear relationships between input and output. Similarly, an increase in the variability in inputs, with no change in the mean, could lead to a change in the mean output as well as change in the extremes. Figure 2.7 illustrates this schematically. The degree of distortion will depend on the characteristics of the hydrological system.

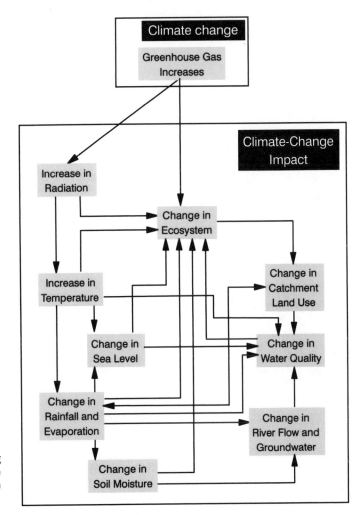

FIGURE 2.6   Impact of increasing greenhouse gas concentrations on the natural hydrological system (Arnell, 1994)

Change in precipitation

Precipitation is the major driving force of the hydrological system. Changes in the amount, intensity, duration and timing during the year will all affect river flows and groundwater recharge, but to what degree will depend on the amount of

1. Change in the mean precipitation

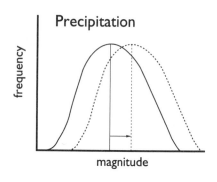

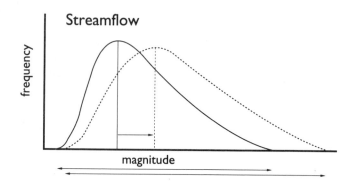

2. Change in the variance of precipitation

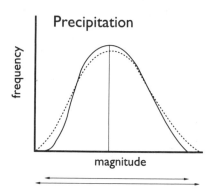

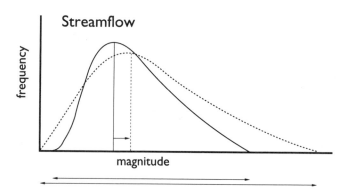

FIGURE 2.7 Schematic illustration of the effects of changing the mean and variance of climatic inputs on the distribution of hydrological output

change and the type of catchment. For example, changes in short-duration rainfall characteristics will have a large effect on flood regimes in highly-responsive catchments, but less impact in large, unresponsive catchments, which will be more affected by changes in the rate of occurrence of prolonged wet spells. An increase in winter temperature would reduce the frequency of snowfall and shorten the season over which precipitation falls as snow.

Change in evaporation

Potential evapotranspiration is the evaporation and transpiration that would occur from an extensive short grass crop with an unlimited supply of water. It is controlled by meteorological demand — as determined by inputs of net radiation, the ability of the air to hold moisture and the rate of renewal of air above the evaporating surface — and plant physiological properties, which include aerodynamic resistance and stomatal conductance. Aerodynamic resistance affects the rate of passage of air across the plant, whilst stomatal conductance controls the rate at which water can be transpired from the plant leaf. Both are dependent on plant characteristics, such as leaf area index. Stomatal conductance is also related to meteorological demand for water: there is an interactive feedback.

Global warming will alter potential evaporation. The most immediate effect will be an increase in the air's ability to absorb water as temperature rises: Budyko (1982) estimated that potential evapotranspiration would increase by 4% for every degree Celsius increase in temperature. However, the effects of increased temperature are complicated by changes in net radiation, humidity and windspeed. Increased cloudiness would be expected to reduce net radiation and increase absolute humidity, so reducing evaporative demand, but with higher temperatures an increase in absolute humidity might still result in a reduction in *relative* humidity, and hence an increase in evaporative demand. The precise effects of global warming on potential evaporation will depend on the current climatic characteristics of a site, as well as details of the meteorological changes.

Vegetation characteristics can also be expected to change as a result of global warming, leading to a change in the rate of potential evapotranspiration. Alterations in climatic regimes will affect the timing and amount of plant growth and the mix of vegetation within a catchment, and plant characteristics may also change as a result of increasing levels of atmospheric CO_2. There is a growing body of research into potential changes in plant physiology caused by altered climate and CO_2 concentrations (Drake, 1992; Kimball *et al.*, 1993; Tyree and Alexander, 1993). This uses controlled laboratory experiments, and increasingly also field experiments with controlled free-air CO_2 concentrations (Hendry *et al.*, 1993).

There are two main types of change in plant physiology. First, experimental evidence shows that stomatal conductance of some plants reduces as CO_2 increases, resulting in a reduction in transpiration. Second, some plants grow more vigorously in a CO_2-rich environment, with the effect depending on how the plant actually absorbs CO_2. Most UK plant species fall into the C_3 group, which are the most sensitive to increased CO_2 concentrations. Increased plant growth caused by global warming might counteract reduced stomatal conductance, and experiments conducted by Gifford (1988) found that plants appeared to maintain some optimum water use. It is, however, very difficult to extrapolate results from controlled experiments, particularly those done in the laboratory, to the catchment. Plant response to elevated CO_2 concentrations may be strongly affected by the presence of adequate nutrients as well as by associated changes in temperature and precipitation, and it is also possible that plants may acclimatise to higher CO_2 concentrations.

Actual evaporation is limited by the amount of water available, and changes in actual evaporation therefore depend not just on changes in potential evaporation, but also on increases or decreases in rainfall. It is possible, if rainfall falls significantly, for actual evaporation to decline while potential evaporation increases.

Runoff generation

In humid temperate areas, runoff is usually generated by water infiltrating into and flowing through soil, and by rain falling on saturated areas. Soil properties influence the rate at which rainfall infiltrates and moves through the soil, and also the time during which soil is saturated. Global warming may affect soil properties, and thus affect the runoff generation process. Higher temperatures and higher rainfall would lead to a loss of soil organic matter, and hence a decrease in the ability of the soil to hold moisture. Increased desiccation in summer would enhance soil cracking, and hence lead to greater infiltration into the soil. Increased rainfall might lead to the development of gleyed layers within the soil, which limit downward infiltration. Changes in soil properties will occur over a long time period, and their effects will probably be noticeable only in small catchments.

Effects of human activities

Many catchments and rivers are heavily modified by human activities, which will complicate the effect of global warming. The most obvious human interventions are structures and procedures to manage water (particularly reservoirs and inter-basin transfers) and abstractions from and returns to water courses: all have direct impacts, intentional and unintentional, on river flows. Human actions also affect catchment land use, and these too can have significant effects on river flow regimes and water quality. Land use decisions may be affected by global warming: farmers may alter cropping patterns or crop mixes (or at the extreme, abandon land or cultivate new land), different amounts of agricultural chemicals may be applied, and a policy of afforestation for carbon sequestration would affect catchment water balances. Of course, human activities are affected by factors other than climate change, and a multitude of human decisions (deliberate and inadvertent) will affect the way an increase in greenhouse gas concentrations impacts upon the water environment and water management.

CAN THE EFFECTS OF GLOBAL WARMING BE SEEN?

This chapter began by introducing the concept of climatic variability over a range of time scales. Both the national temperature record (Figure 2.2) and the global record (Figure 1.1) show an increase in temperature in recent years, and most of the warmest years on record have occurred during the 1980s and 1990s. Is this evidence of global warming?

There are several difficulties in attributing any observed climate change to global warming. First, it is very difficult to separate any global warming signal from the considerable year-to-year variability in climate, because the signal-to-noise ratio will be very low. A further complication arises from periodic fluctuations in current climate that are due to such features as El Niño/Southern Oscillation, which are particularly strong in certain parts of the world. It would be necessary to develop a very sophisticated test of trend, which would not assume that the current climate is stationary, and which would be able to detect non-linear as well as linear trends. The second major problem is that different regions may show very different variations over time: some parts of the world might experience a warming in a certain time period, while others might be cooling.

A more sophisticated approach to the detection of global warming takes the so-called "fingerprint" approach. Global warming caused by an enhanced greenhouse effect will, if it occurs, have a particular fingerprint, which will differ from the effects of natural climate variability or regional climate changes. This fingerprint would include, for example, an increase in global temperature at the earth's surface and a decrease in temperature higher in the atmosphere. The possible regional pattern of change can be simulated using climate models, and although there is considerable variability between models, there are quite consistent projections of a more rapid increase in land than sea temperatures, and a greater temperature rise in the Northern Hemisphere. A comprehensive fingerprint test would therefore look at many dimensions of climate at global and regional scales and compare observed changes with those expected under the global warming theory. This, of course, requires a comprehensive global climate monitoring system measuring the right variables, and the World Meteorological Organisation (WMO) is currently developing the Global Climate Observing System (GCOS: Spence and Townshend, 1995; Karl *et al.*, 1995). A special issue of the journal *Climatic Change* outlines proposals and methodologies for monitoring many dimensions of climate, including not only temperature but variables such as clouds (Rossow and Cairns, 1995), the cryosphere (Walsh, 1995) and land cover (Ropelewski, 1995). There are, however, many problems with defining consistent long time series of many variables, including measurement error (Groisman and Legates, 1995) and the scope of the observing network (Jones, 1995).

A significant omission from the variables covered in the special issue of *Climatic Change* is river flow. This is rather surprising, because river flow has the potential to integrate data over a region, and because changes in precipitation tend to be amplified in changes in runoff (as will become clear later in the book). The major problems, however, are that the records are rarely more than a few decades long, that river flows in many catchments with long records are affected by human interventions such as land use change and reservoir construction, and that the effects of features such as El Niño/Southern Oscillation can be seen on river flows in some parts of the world — for example in the western United States (Kahya and Dracup, 1994) and South America (Mechoso and Iribarren, 1992). However, there have been several attempts to identify trends in river flow data using a variety of statistical techniques (Schädler, 1987; Kite, 1989; 1993a; Chiew and McMahon, 1993; Burn, 1994; Lettenmaier *et al.*, 1994; Lins and Michaels, 1994; Mitosek, 1995; Hisdal *et al.*, 1995; Marengo, 1995).

Chiew and McMahon (1993), for example, applied five statistical trend tests to data from 30 Australian catchments with minimal human intervention, and found no clear evidence for any trend caused by global warming. Kite (1993a) found some linear trends in streamflow data from Canada, but there were no clear spatial patterns in the trends. Marengo (1995) also found no clear evidence of trend in South American streamflow data. In a global study, using data from many different regions, Mitosek (1994) identified a number of trends in the mean and variance of flows, but could not identify consistent patterns or attribute trends to any cause. Hisdal *et al.* (1995) looked for trends and jumps in hydrological time series from the Nordic region, and although they found examples of both and a few regionally-consistent patterns, they too could not prove that they were due to climate change.

In contrast, Lettenmaier *et al.* (1994) detected clear increases in November to April streamflow across much of the United States over the period 1948 to 1988, especially in the north-central United States. Lins and Michaels (1994) explicitly related regional and seasonal increases in streamflow to global warming, using a longer period of record. In west-central Canada, Burn (1994) found a greater number of rivers exhibiting earlier spring runoff than could be attributed to chance, a finding that is consistent with the effects of global warming.

The evidence for global warming having a noticeable effect on hydrological behaviour is not yet convincing, but it does seem to be accumulating. How, then, is it possible to estimate the impacts of future climate change on hydrological regimes, and what might these changes mean for water resources? The remainder of this book aims to provide some answers.

CHAPTER 3

Assessing the effects of climate change

Climate change impact assessments are estimates of what might happen under specified scenarios. As such, they must be based on a rigorous, well-documented methodology, and all the stages in the process must be credible and scientifically-supportable. Assessments follow one of three basic methods (Kates, 1985; Carter *et al.*, 1994).

- The *impact* approach is essentially a straightforward linear analysis of cause and effect: if climate were to change in a defined way, what would happen? Virtually all hydrological and water resources impact studies have followed this approach.
- The *interaction* approach allows for the effect of other changes and of feedback between climate and the system affected by climate change. Studies investigating the relative importance of climate and land use changes on river flows would fall into this category.
- The *integrated* approach looks at society as a whole, considering the hierarchies of interactions within each sector and the interactions and feedback between sectors (Rotmans, 1990). An example at the catchment scale is the Mackenzie Basin Integrated Study (MBIS) which is examining in an integrated manner potential impacts of global warming on different sectors of the economy, landscape and society in the Mackenzie Basin in northern Canada (Cohen, 1995).

IMPACT ASSESSMENT METHODOLOGY

A general framework for conducting a climate change impact assessment was prepared, under the auspices of the IPCC, by Carter *et al.* (1994). They defined seven steps as shown in Figure 3.1. The steps are consecutive, but can be repeated.

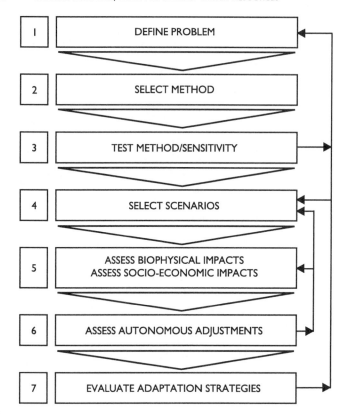

FIGURE 3.1 Seven steps in climate impact assessment (Carter *et al.*, 1994)

The first step, rather obviously, is to define the problem in terms of impact of interest (e.g. hydrological regime or water resource impact), geographic location and time frame.

The second step is to select an analysis method. In some impact areas it is possible to conduct a conventional experiment. For example plant physiological responses to increased atmospheric CO_2 concentration can be determined experimentally although it may be difficult to ensure that the experiment realistically reproduces natural conditions. Experimentation has also been widely used to investigate the effects of land use change on catchment hydrological regimes (e.g. by clear-felling a forest cover), but is difficult to apply when investigating the effects of changes in climatic inputs. The CLIMEX experiment (Jenkins *et al.*, 1995) is attempting to measure the effects of a change in CO_2 concentrations on hydrological regimes in a small forested catchment in southern Norway entirely enclosed within a greenhouse, but experimentation is not widely applicable in hydrological climate change studies. Instead, a range of impact projection and analogue methods have been adopted. Table 3.1 summarises methods applied in hydrological impact assessments, and gives a few examples.

TABLE 3.1 Methods for hydrological impact analysis, with some example studies

	Direct estimation of hydrological impacts	Application of changes in climate inputs to hydrological models
Arbitrary changes		Nemec & Schaake (1982)
Temporal analogue	Krasovskaia & Gottschalk (1992); Frederick (1993)	Palutikof (1987)
Spatial analogue	Arnell et al. (1990)	
Climate model simulations	Miller and Russell (1992); Smith and Tirpak (1989)	Bultot et al. (1988a); Cohen (1987; 1991) and many others

One dimension of the classification represents the origin of the information on possible climate change, whilst the second distinguishes between direct estimation of hydrological changes and estimation of the hydrological implications of climatic change inputs using a hydrological model.

Most studies fall into the bottom right box because they use hydrological models to determine the hydrological implications of changes in climate as simulated by numerical climate models.

Step three is the validation of the method, which is necessary to ensure the credibility of the results, and step four is the definition of climate change scenarios. The final three steps are concerned with estimating the impacts and the effect of any adjustments that might take place independently of climate change but which might mitigate climate change impacts, and the evaluation of deliberate adaptation strategies (these are defined and discussed at the end of the chapter).

In hydrological and water resources impact studies the critical components of this process are the selection of a methodology and the definition of climate change scenarios.

CLIMATE CHANGE SCENARIOS

Climate change impact assessments compare the conditions that might be expected in the absence of climate change with those that might exist with climate change. Both these states are described by a scenario, which can be defined as "a coherent, internally-consistent and plausible description of a possible future state of the world" (Carter et al., 1994 p15). The baseline represents current conditions against which projections may be compared and to which scenarios can be applied.

Three interrelated sets of controls determine current hydrological regimes and water resources in a catchment, and need to be considered in impact assessments (Figure 3.2). Climate defines the inputs of precipitation, temperature and evaporative demand into the catchment, whilst land use defines the vegetation and land surface cover within the catchment. The water management system determines how the resources in the catchment are exploited. Catchment physiography also determines hydrological regimes and causes differences between catchments, but can be regarded as constant over time. Baseline conditions need to be established for all three components and scenarios defined for each. In practice, virtually all hydrological and water resources impact studies have assumed land use and the water management system will remain constant over time, and have considered only scenarios for changes in catchment climatic inputs. One exception is Kite's (1993b) study, which considered effects on flow regimes of changes in climate and land use, using a hydrological model with parameters based on land cover type.

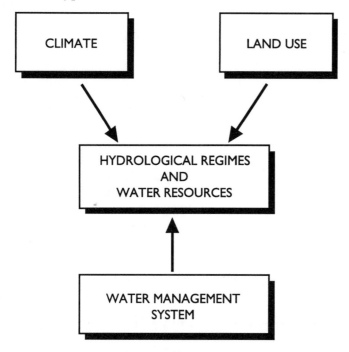

FIGURE 3.2   Three controls on hydrological regimes and water resources

Global warming climate change scenarios include (implicitly or explicitly) two components. The first is an emissions scenario, which determines the concentration of greenhouse gases in the atmosphere at a given time, and the second is a climate scenario which strictly defines the changes in climate (temperature, rainfall etc.) for a given change in greenhouse gas concentrations. Emissions scenarios make assumptions about the rate of change in human activity and the rate at which natural land, ocean and atmosphere systems can absorb greenhouse gases. In 1990, the IPCC defined four emissions scenarios, one representing a "business-as-usual" case assuming the continuation of present population and economic growth trends and no explicit response to global warming (SA90), and the remaining three representing different policies for restricting the emission of greenhouse gases (IPCC, 1990a). In 1992 the IPCC introduced six "business-as-usual" scenarios, each making different assumptions about population growth and economic development but all assuming no explicit response to global warming above that already adopted (Figure 3.3). Studies into greenhouse gas mitigation costs and policies examine a range of emissions scenarios but virtually all impact assessments based on climate model simulations have implicitly used a "business-as-usual" scenario: this is what climate model simulation runs are based on. Kwadijk and Middelkoop (1994) and Kwadijk and Rotmans (1995), however, explicitly compared the hydrological impacts on river flows in the Rhine basin under the 1990 business-as-usual emissions scenario (SA90) with those under a scenario of maximum effort in greenhouse gas mitigation. These studies represent the only published assessment of the impact of different emissions policies.

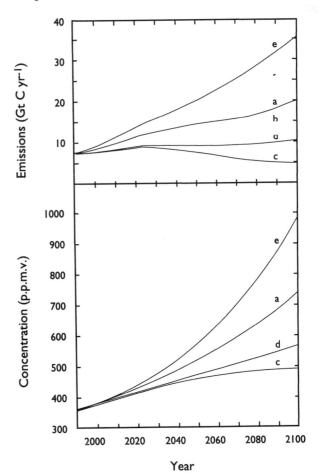

FIGURE 3.3 Atmospheric CO_2 concentrations under four of the IPCC (Wigley and Raper, 1992) emissions scenarios

Land use scenarios can reflect influences other than climate change — such as urbanisation and changes in agricultural demand, practices or policies — but can also incorporate effects of global warming. These effects may be direct, as when a change in climate would cause a change in catchment vegetation cover or agricultural potential. The effects may also be indirect, whereby land use changes in a catchment in response to effects of climate change elsewhere; an example might be a change in agricultural use following a change in produce prices (shifts in world grain prices might affect the decision of individual farmers in individual catchments to grow wheat rather than, for example, graze cattle). Such effects are of course difficult to estimate (Riebsame *et al.*, 1994), and land use scenarios accounting for climate change will be very uncertain.

Water management scenarios define the change in objectives, pressures and demands placed on water resource systems over time. Changes in demands (for water supply, power and recreation) occur as a result of population change, economic development and restructuring, and the use of more efficient appliances and production techniques. Many techniques have been developed to forecast demands of different types, but forecasts are acknowledged to be very uncertain (Stakhiv, 1993). It is more difficult to forecast changes in management objectives over time. Examples might include the introduction of water quality standards or an increasing concern with environmental issues.

Current conditions — against which changes are assessed — are defined by the baseline so it is important that this baseline is clearly stated; a different baseline may mean a different estimate of change. The climatic baseline period is most important. If the aim of the study were to estimate the total effects of human-induced climate change, the baseline climate period ought to date from pre-industrial times (before 1750).

In practice, few studies take such a perspective, and very few quantitative data are available. Studies therefore examine the potential change from "present" conditions (which might already incorporate some greenhouse warming). The current World Meteorological Organization (WMO) standard climatic period (termed the climatic normal) runs from 1961 to 1990, but this includes the decade of the 1980s which contains some of the warmest years over the last 100 years. The previous standard period (from 1931 to 1960), however, contains too few data. Most studies use the entire period of record available as the baseline but this varies from study to study and, within a study, from catchment to catchment. Other studies have therefore used defined 30-year periods (or shorter): the period 1951 to 1980 is assumed by many climate modelling groups to define - current - conditions (initialisation data and fixed boundary conditions come from this period), and global temperature figures are expressed as departures from the 1951 to 1980 average (see Figure 1.1). Figure 3.4 summarises baseline periods used in several hydrological climate change impact studies, and emphasises the variety of periods used.

Table 3.1 showed four ways of defining climate change scenarios: arbitrary change, temporal analogues, spatial analogues and climate model simulations. These are now outlined in some detail, with greatest attention directed towards the use of climate model output.

Arbitrary changes in climate inputs

Studies based on arbitrary change scenarios investigate the implications of specified changes in hydrological regimes on a system (the first column of Table 3.1), or determine the sensitivity of the hydrological system to changes in climatic inputs (the second column). The approach is basically a sensitivity analysis, examining the sensitivity of a catchment or water resources system to change in inputs. It provides valuable information about vulnerability but does not provide a basis for the estimation of possible future conditions.

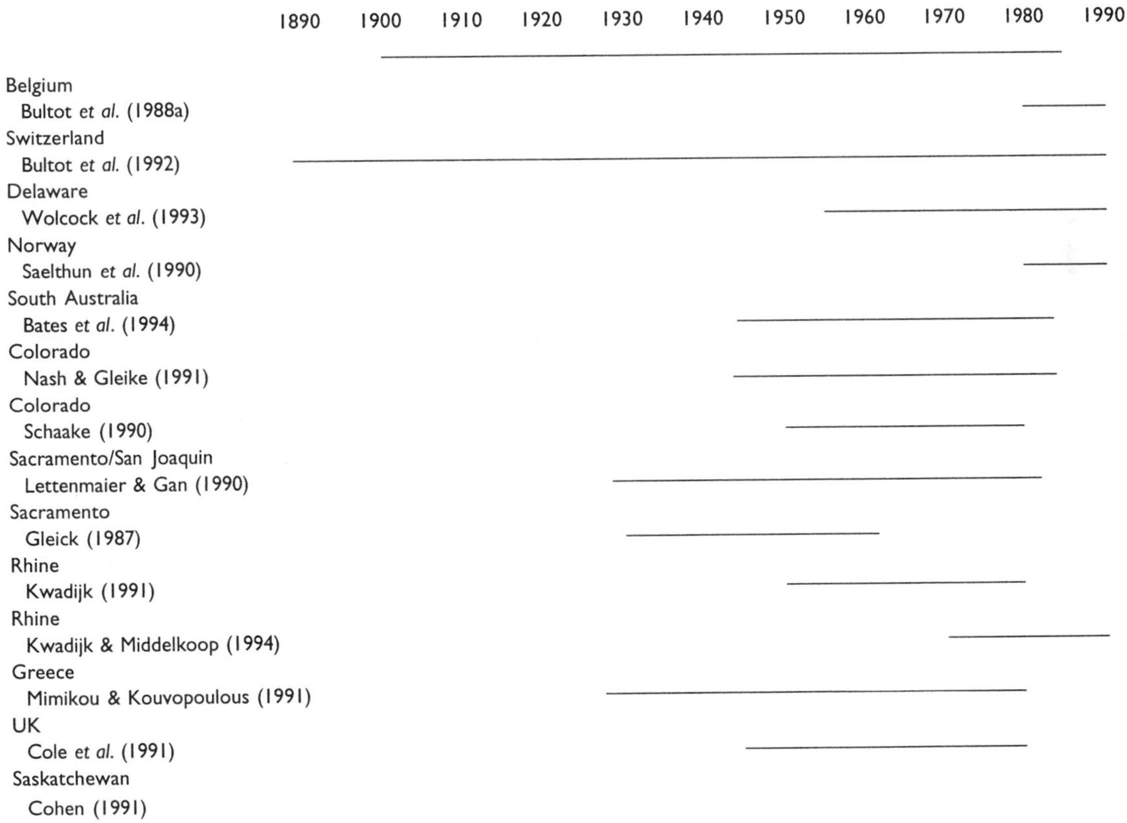

	1890	1900	1910	1920	1930	1940	1950	1960	1970	1980	1990

Belgium
  Bultot et al. (1988a)
Switzerland
  Bultot et al. (1992)
Delaware
  Wolcock et al. (1993)
Norway
  Saelthun et al. (1990)
South Australia
  Bates et al. (1994)
Colorado
  Nash & Gleike (1991)
Colorado
  Schaake (1990)
Sacramento/San Joaquin
  Lettenmaier & Gan (1990)
Sacramento
  Gleick (1987)
Rhine
  Kwadijk (1991)
Rhine
  Kwadijk & Middelkoop (1994)
Greece
  Mimikou & Kouvopoulous (1991)
UK
  Cole et al. (1991)
Saskatchewan
  Cohen (1991)

FIGURE 3.4 Baseline periods used in hydrological and water resource impact studies

Many hydrological and water resource impact studies have examined the implications of arbitrary changes in climatic and hydrological characteristics. Nemec and Schaake (1982) in one of the earliest impact studies examined change in annual runoff and reservoir reliability for two study catchments, for defined changes in rainfall and temperature (Figure 3.5a). Schaake (1990) subsequently used a water balance model to simulate the spatial effect across south-east USA of defined changes in precipitation and potential evaporation (Figure 3.5b). Mimikou et al. (1991) calculated change in the risk of a Greek reservoir system being unable to supply the design volume of water, and Wolock et al. (1993) tabulated changes in the risk of the Delaware River basin reservoir system being unable to supply the water required by New York City. There are many other examples. Such studies can provide useful information about vulnerability and sensitivity to change but are not really assessments of the potential impact of climate change.

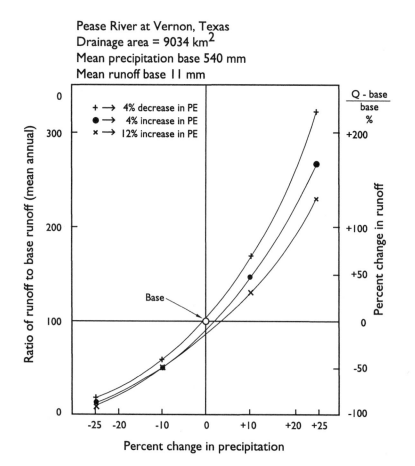

FIGURE 3.5a Change in annual runoff with change in precipitation and potential evaporation: Pease River, Texas. Nemec and Schaake (1982)

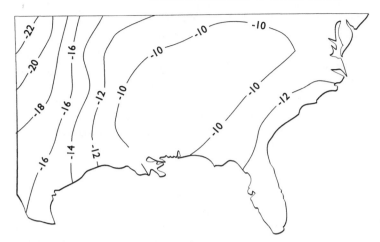

FIGURE 3.5b Change in annual runoff across the south-east United States: no change in precipitation, potential evaporation increased by 10% (Schaake, 1990)

Temporal analogues

There are two types of temporal analogue that use information from the past as a scenario for future changes. The first uses data from the instrumental record. Krasovskaia and Gottschalk (1992), for example, compared river flow regime types in the Nordic countries in warm years and cool years, showing a reduction in the importance of snowfall and snowmelt from the cool to the warm years. Palutikof (1987) compared simulated river flows in several British catchments in warm (1934–1953) and cool (1901–1920) periods, finding a shift in both the magnitude of flow and the timing of flow through the year (Figure 3.6).

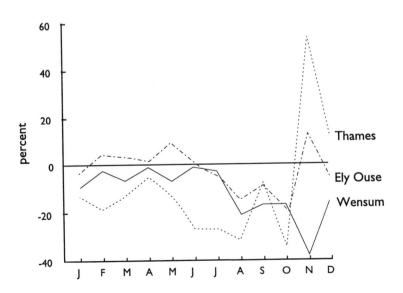

FIGURE 3.6 Monthly difference between warm and cold periods in catchments in eastern England (Palutikof, 1987)

There are a number of problems with the approach. First, and most importantly, it is assumed that differences between warm and cool periods in the past represent a reasonable analogue for the difference between current "cool" conditions and future "warm" conditions following global warming, even though the factors causing the difference are not the same. In fact, past differences may be due to random fluctuations, not some systematic forcing: they cannot therefore be expected to have the same signal as the effects of an increasing concentration of greenhouse gases in the atmosphere. Second, different analogues may be chosen using different climatic data sets. The warmest decade on record will differ depending on whether local, regional, hemispherical or global temperature data are used. Considering the Northern Hemisphere as a whole, the 20-year warm period 1934–1953 was 0.5°C warmer than the 1901–1920 cool period; in Europe, 1934–1953 was cooler than 1901–1920 (Palutikof *et al.*, 1984). Third, the difference in temperature between warm and cool periods in the instrumental record may not be as large as the increase in temperature anticipated due to global warming. Finally, instrumental records are often very short and it may be difficult to find any differences between warm and cool periods. Nevertheless, information from the past may give important insights into the behaviour of an environmental or water resources system under extreme conditions. An example is the MINK project in Missouri, Iowa, Nebraska and Kansas, USA, which examined how the drought of the 1930s would impact upon the present and future economy and water use of the region (Frederick, 1993).

The second type of temporal analogue uses palaeoclimatic information, derived from reconstructions of past environments based on geological (land forms, sediments and the age of groundwater) and botanical (pollen analysis and dendrochronology) evidence. The Holocene climatic optimum (6.2 to 5.3 ka BP), the last interglacial (125 ka BP) and the Pliocene (3 to 4 ma BP) have been proposed as analogues for global warmings of 1, 2 and 4°C respectively (IPCC, 1990). There are two major assumptions with the use of palaeoclimatic analogues. It is assumed that the relationships between form and process operating today are the same as those operating in the past, and therefore that past processes (river flow, rainfall, etc.) can be inferred from past form. Second, it is assumed that the effects of change in climate are independent of the cause of that change; more specifically, it is assumed that the changes in climate due to changes in solar radiation following changes in the Earth's orbit will be similar to the changes due to increasing concentrations of greenhouse gases. This assumption is not necessarily true (Mitchell, 1990).

A practical problem with the approach is that it is very difficult to obtain quantitative information: the best prospects lie with flood reconstruction. Knox (1993), for example, reconstructed a flood chronology for tributaries of the Upper Mississippi from sedimentological evidence, and showed large changes in flood risk for modest changes in climate (Figure 3.7).

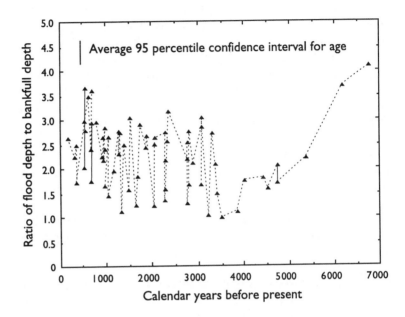

FIGURE 3.7 Flood magnitudes in south-west Wisconsin during the Holocene (Knox, 1993). The vertical axis shows the ratio of flood magnitude to bankfull depths: a value of 2 corresponds to a flood with a return period at present of between 30 and 50 years.

Palaeohydrological reconstructions are probably of greatest value in indicating how the fluvial system responds to changes in climatic and other conditions, rather than as explicit analogues for the future (Arnell, 1996). They can provide data for the validation of models which attempt to describe the effects of change.

Spatial analogues

The spatial analogue approach assumes that the future climate of a region can be represented by the present climate of another. Parry *et al.* (1988) used a number of spatial analogues to estimate possible changes in agriculture; for example the future climate of Iceland was assumed to be like the current climate of Scotland, and the current climate of the US mid-west was assumed to be an analogue for the Canadian mid-west in the future. However, local and regional climate is a result of the interaction of large-scale circulation features with local characteristics, such as topography and proximity of ocean, and it is not always realistic to transfer climate from one region to another. It is even more difficult to transfer hydrological information from one region to another, because hydrological regimes are a function of both climate and the physical attributes of the catchment. Arnell *et al.*

(1990) attempted to use south-western France as an analogue for future conditions in southern Britain but found that differences in geology and topography made it impossible to transfer French hydrological characteristics to England with any credibility.

Empirical relationships predicting hydrological properties from catchment climatic and physical characteristics can be seen as spatial analogues: they are substituting variation in space (between catchments) for variation over time. Examples include the regression equations used to predict mean annual flood from catchment area, rainfall, soil characteristics and channel slope as developed in the UK Flood Studies Report (NERC, 1975), empirical equations predicting low flow properties in Europe from catchment rainfall and soil characteristics (Gustard *et al.*, 1992), and the relationships developed by Langbein (1949) predicting average annual runoff in the USA from average annual precipitation and temperature. Many such empirical relationships have been developed in hydrological environments. In principle, they can be used to estimate the implications of a change in climate on hydrological characteristics, and the Langbein relationships have been used in this way by Revelle and Waggoner (1983). Brown and Katz (1995) used regional relationships to define spatial analogues for temperature extremes in the United States. In practice, however, the estimated sensitivity to change in climate is dependent both on the form of the model and the values of the empirical regression coefficients, which are themselves dependent on the data used and variables included. Empirical models also internalise relationships between the dependent hydrological characteristic and variables not included in the empirical model. The value of the coefficient relating annual rainfall to annual runoff, for example, implicitly assumes a particular relationship between rainfall and evaporation. Schaake (1990) remarked that Revelle and Waggoner's (1983) estimates of sensitivity of annual runoff in the western USA to change in temperature and precipitation were quantitatively different to those based on model simulation studies, and Arnell (1992) illustrated the large differences in implied sensitivity of annual runoff in UK catchments between regression relationships relating runoff to rainfall and potential evaporation.

Scenarios based on climate models

The most convincing way to estimate the effects of a change in the factors affecting climate (solar radiation, sea surface temperatures, land cover or greenhouse gas concentrations) is to use a numerical atmospheric simulation model. General circulation models (GCMs) solve the equations of conservation of momentum, mass and energy for an air parcel, at all points in space and over time, in order to simulate energy and moisture fluxes in the atmosphere and hence the growth, development and decline of large-scale atmospheric circulation features. Current GCMs not only simulate behaviour within the atmosphere but also represent the relationships and fluxes between the atmosphere, the land surface, and the ocean. Many GCM experiments have been run to investigate the effects of increasing greenhouse gas concentrations, showing a rise in global mean temperature of between 1.5°C and 4.5°C following a doubling of equivalent atmospheric CO_2 concentrations, and considerable spatial variability in temperature and other climatic changes (IPCC, 1990a; 1992). Experiments have considered the equilibrium climate of an atmosphere with a CO_2 concentration stabilised at twice pre-industrial levels (as reported in IPCC, 1990a), as well as the transient evolution of climate as greenhouse gas concentrations gradually increase (IPCC, 1992; Murphy and Mitchell, 1995).

The simplest GCM-based climate change scenario directly takes the changes in precipitation, temperature or runoff as simulated by a GCM. Miller and Russell (1992), for example, determined the change in annual runoff for 33 of the world's major rivers directly from the output of the Goddard Institute for Space Studies (GISS) GCM, and the US Environmental Protection Agency examined the effects of changes in river flows in the United States as determined directly from GCMs (Smith and Tirpak, 1989).

This approach is not appropriate for hydrological and water resources impact assessments of specific catchments, for two main reasons:

- the GCM may not simulate the climate in the study region particularly well, and

- the spatial scale of current GCMs is too coarse for hydrological application.

The spatial resolution of global GCMs is about 3.75° × 2.5° (or approximately 80 000 km² at 50° N), whilst hydrological studies ideally need information at a scale of 10^3 km² or finer. The spatial resolution of GCMs is limited by available computer resources and the coarse resolution is one of the reasons why GCMs do not necessarily simulate local climate very well. Meso-scale patterns such as depressions, which operate at the scale of a few tens to a few hundreds of kilometres, cannot be simulated explicitly in coarse-resolution GCMs but they may have very important effects on local and regional climate.

Catchment hydrological processes operate at a daily, or finer, time step, and ideally hydrological impact assessments require information on changes at this time scale. GCMs work at a time resolution of minutes but their coarse spatial resolution means that this high time resolution is not particularly helpful; the short-duration fluxes are calculated over a very large geographic area.

It is, however, possible to develop climate change scenarios from climate models and a variety of methods of varying degrees of sophistication have been developed. There are essentially two key issues: how to downscale from the GCM resolution to the catchment scale, and whether to use the GCM climate simulations directly or to define changes in climate which may be applied to historical data. When using GCM simulations of temperature, precipitation and other meteorological characteristics as direct inputs to hydrological models it is assumed that the GCM simulates climate well; when using the GCM to define changes in climate, it is assumed that the GCM estimates of change are reliable even if the simulated current climate is not. Changes in temperature and precipitation (and other meteorological variables) can be applied by simply perturbing historical time series at the location of interest, or by altering the parameters of a stochastic weather generation model based on historical experience (Wilks, 1992; Bates *et al.*, 1994; Racsko *et al.*, 1991). The main disadvantage of perturbing an historical time series is that it is difficult to change the observed temporal structure (such as the length of dry spells and inter-annual variability). The main problem with using a stochastic weather generator is to develop a stochastic model which reproduces accurately the characteristics of observed data and has parameters which can be perturbed from GCM output.

There are several ways to downscale climate model output to the spatial resolution required in catchment hydrological studies. The simplest approach is simply to interpolate climate model output down to the finer spatial scale required. The interpolation may be subjective, e.g. using the nearest grid point, or may use objective statistical interpolation techniques. Most hydrological studies have used relatively simple interpolation techniques to determine scenarios for individual catchments, and most have created scenarios for changes in monthly mean climate which can be applied to catchment data. Cohen (1991), for example, interpolated changes in mean monthly temperature and precipitation from five GCMs onto a consistent set of seven $4° × 5°$ grid points across the Saskatchewan River basin in Canada. Daily time series can be perturbed by altering daily data according to the monthly changes (e.g. Bultot *et al.*, 1988a; 1992). Bates *et al.* (1994) determined changes in the parameters of a daily stochastic weather generation model for a catchment in South Australia from changes in daily precipitation, humidity, temperature and net radiation as simulated by the CSIRO9 GCM at the grid point nearest the catchment.

A variant on this approach introduces some climatic or

meteorological expertise into the interpolation of GCM output and the definition of change scenarios. Robock *et al.* (1993) proposed combining GCM output with knowledge of regional climate to produce change scenarios for several regions. Their scenarios for sub-Saharan Africa took account of the observation that in dry years the rainy season tended to begin later rainstorms were less intense and, in some parts of the region, less frequent. Figure 3.9 illustrates their procedure for generating a daily rainfall scenario for the Sahel when rainfall increases; the start of the rainy season is moved forward by α days, and a corresponding number of days is duplicated in the middle of the rainy season. Rainfall magnitudes are increased by a factor γ.

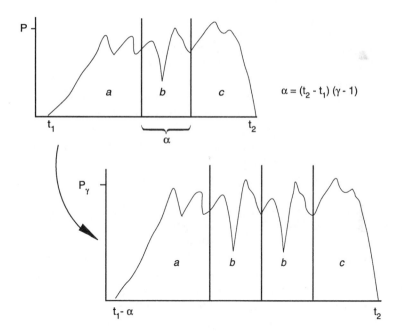

FIGURE 3.9 Creating a scenario for increased rainfall in sub-Saharan Africa (Robock *et al.*, 1993)

Downscaling using empirical relationships between large-scale climate and point climate

This approach assumes that large-scale atmospheric features, at the scale of several thousands of kilometres, are well simulated by the climate model, and that stable relationships exist between these features and local detail. There are several variants to the approach. Kim *et al.* (1984) estimated point monthly rainfall and temperature at a number of locations in Oregon from the state-wide average monthly values, by first calculating empirical orthogonal functions (EOFs) describing the spatial variability across Oregon in monthly climate. A summary index characterising the year-to-year variability of the first EOF (which explained most of the observed variance) was found to be very strongly correlated with inter-annual variability in state-wide climate; given a value of large-scale temperature, for example, it was then possible to calculate the summary index for the first EOF, and then determine point temperature by rescaling the point loadings on the EOF. A rather simpler approach to estimate point monthly rainfall and temperature in Oregon was presented by Wigley *et al.* (1990). They developed a suite of regression equations predicting, for each point and month, temperature (or precipitation) from large-scale average temperature (or precipitation), pressure in central Oregon and pressure gradients across the state. There was considerable spatial variability in the strength of correlation, meaning that in some areas it was difficult to predict local climate from large-scale climate. Local temperature was found to be easier to determine from large-scale climate than local precipitation.

Both Kim *et al.* (1984) and Wigley *et al.* (1990) were concerned with estimating point monthly temperature and precipitation. Their methods could be applied to determine point monthly climate under current and $2 \times CO_2$ condition, and either these values could be used to run a hydrological model, or the monthly changes could be applied to an observed time series. Karl *et al.* (1990) described a far more complicated procedure for estimating daily temperature and precipitation at a point, based on a development of methods used for weather forecasting. The procedure essentially develops relationships between free atmospheric variables (such as pressure, wind components and humidity) and surface observations (temperature and precipitation), using the framework shown in Figure 3.10. The first stage uses principal components analysis to identify the combination of the 22 free atmosphere variables considered that show the greatest climatic signal, and the second stage uses canonical correlation analysis to develop summary variables that show the greatest correlation between free atmospheric variability and surface observations. The resulting canonical variables are then related to surface observations using inflated regression analysis (as an alternative to least squares regression; it minimises the differences between the statistical distributions of predictor and predictand). Karl *et al.* (1990) applied the method to estimate from GCM output daily temperature, precipitation and cloudiness at five sites in the United States. They found the best results when the free atmospheric variables were derived directly from GCM output, rather than from GCM output restandardised by point free atmospheric observations. Although the method simulated local climate well from large-scale data, it requires a considerable amount of analysis effort.

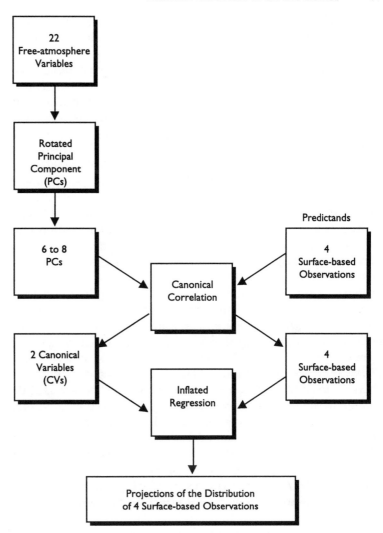

FIGURE 3.10 Flow chart describing the development of relationships predicting surface observations from free atmosphere variables (Karl et al., 1990)

Another way of relating large-scale climatic features to point climate uses discrete circulation pattern types. The basic method has three stages; define circulation pattern types (westerly, anticyclonic and so on), determine point precipitation characteristics for each type (studies so far have focused on precipitation; other variables such as temperature could also be analysed) and then, for a daily time series of circulation pattern types, simulate daily precipitation. Examples include studies by Wilby (1995) in central England, Hay et al. (1991; 1992) in the Delaware Basin, eastern USA, Bardossy and Plate (1992) in the Ruhr Basin, Germany, and Wilson et al. (1992) in Washington State, USA. Time series of circulation pattern types could be taken directly from GCM output and converted to daily point precipitation but most studies have generated time series of circulation pattern types stochastically with model parameters derived from GCM or observed output.

Some studies using circulation pattern types have developed independent relationships for each point of interest (e.g. Hay *et al.*, 1991). Summed over time, the correct spatial pattern appears but there is no spatial structure in individual precipitation events. This is only a problem if spatial variability within a catchment is important for the hydrological or water resources simulation model (if, for example, the hydrological model needs distributed rainfall inputs or a water resources system model needs to know timings of inputs into several reservoirs in different catchments). However, Bardossy and Plate (1992) and Wilson *et al.* (1992) both incorporated spatial correlation into their stochastic precipitation routines, enabling them to generate daily precipitation, for a given circulation type, with a realistic spatial structure.

In principle, the method can be used to define climate change scenarios by generating time series of circulation patterns under altered climatic conditions, and assuming that the relationships between circulation pattern type and precipitation characteristics remain constant. However, Hay *et al.* (1992), studying the Delaware Basin, found that whilst the climate model simulated changes in precipitation in a warmer climate, there was little change in the frequency of weather types: they concluded that changes in regional precipitation might reflect changes in the characteristics of given weather types rather than the occurrence of weather types. Matyasovszky *et al.* (1993) incorporated information on pressure heights for each weather type under current and $2 \times CO_2$ climates in their model for eastern Nebraska, and were thus able to consider the effects of changes in weather type properties.

Empirical relationships between large-scale climatic features and point climate can be used to create high-resolution scenarios for hydrological studies but the credibility of these scenarios is dependent on the credibility of the simulated large-scale circulation patterns and the assumption that empirical relationships based on past data will hold under changed conditions; the estimated parameters of these relationships may also vary depending on the period of data used.

Downscaling using a nested regional high resolution model

The fourth method of downscaling uses a regional high resolution climate model nested within the global climate model (Giorgi and Mearns, 1991; Giorgi *et al.*, 1992; 1994; Jones *et al.*, 1995). The global model provides information on boundary conditions to the regional model, which then simulates regional climate in more detail considering such sub-grid detail as topographic features and coastlines. Regional climate models have a spatial resolution of the order of 60 km (or 0.5° × 0.5°), which is fine enough to yield a realistic representation of mid-latitude synoptic systems (Giorgi and Bates, 1989). The nesting is one-way, because the regional model does not pass information back to the global model. The definition of the domain of the regional model is important. It must not be so small that there is no space for regional features to develop, but it cannot be so large that it develops large-scale atmospheric features inconsistent with the driving global model. Giorgi *et al.* (1994) showed that a nested regional model produced a more realistic simulation of precipitation over the United States than the driving global model alone and also that estimated changes in climate were different: precipitation changes differed locally in magnitude, sign, and spatial and seasonal detail, as the higher-resolution regional model simulated local circulation features in more detail. McGregor and Walsh (1994) used a model with multiple nesting down to 60 km to simulate changes in climate in Tasmania. They found that changes in precipitation were related primarily to modifications to low-level synoptic fields, which were not resolved at the coarse GCM resolution.

Global climate models operate on a very short time step, but their coarse spatial resolution means that it is difficult to extract meaningful information about daily variability. However, nested models operating at resolutions of the order of 50 × 50 km can provide useful information on day-to-day variability and the changes that might occur due to increased greenhouse gas concentrations. Mearns *et al.* (1994), for example, analysed daily output from Giorgi *et al.*'s (1994) nested model and found a general increase in the variability of daily precipitation. In some locations, e.g. coastal north-west America, there was little change in daily rainfall frequency, but changes in mean intensity. In other areas, e.g. the central Great Plains, there were more significant changes in the frequency of rainfall. Mearns and Rosenzweig (1994) then used the simulated changes in daily rainfall to alter the parameters of a daily stochastic rainfall and weather generation model, and simulated changes in crop yield. They found significantly different changes in yield when they accounted for changes in day-to-day variability. Mearns and Rosenzweig (1994) did not use the daily rainfall as simulated by the nested model directly: the differences between the observed current daily variability and the simulated current variability were too great.

Scenarios derived from nested model simulations have considerable potential in hydrological impact assessments but have not yet been so used.

Which climate model?

The preceding section outlined methods for downscaling the output from a global climate model to a spatial scale relevant for hydrological impact assessments. But which climate models should be used and what do climate model simulations actually represent? Three issues are important:

- accounting for the different predictive abilities of different climate models,
- using output from equilibrium or transient climate change experiments, and
- defining change scenarios for specific time periods.

Different GCMs simulate current climate with varying degrees of precision. All simulate basic circulation characteristics and the annual cycle reasonably well but the quality of regional simulations varies considerably; the greatest differences lie in precipitation. A GCM that simulates climate well in one region may simulate it very badly in another. The quality of prediction of a GCM is impossible to assess, because there are obviously no observations to compare against. The simplest response to this model uncertainty is to develop scenarios from several models and treat each scenario equally. Another response is to weight each model scenario according to the ability of the model to simulate current climate, although a good simulation of current climate in a region does not necessarily guarantee a reliable prediction of the future regional climate. A third option is to take an average across all the GCMs considered, perhaps weighting by current performance (Hulme *et al.*, 1995a) but the major problem with this approach is that the impact of the average climate change is not necessarily the same as the average impact of each scenario, because of the non-linear relationships between climate and impact.

The earliest GCM experiments (IPCC, 1990a) into the effects of increasing greenhouse gas concentrations compared current conditions with the equilibrium climate that would occur if effective CO_2 concentrations were stabilised at twice current values. More recent experiments (IPCC, 1992) have simulated the effects of a gradual increase in CO_2 concentrations. These transient experiments simulate explicitly the effects of lags in the ocean system on the rate and geographic pattern of changes in climate. In comparison with equilibrium experiments, such transient simulations tend to show a smaller and more delayed increase in temperature around the margins of continents, due to the effects of the oceans. In principle, transient experiments should provide more realistic projections of the rate and pattern of change in climate, but there are several problems with creating scenarios from transient simulations.

The simulated transient climate time series cannot be input directly into impact models, largely because of the coarse spatial resolution, so it is necessary to define changes, relative to a control simulation with stable CO_2 concentrations, to be applied to observed climate time series. One way of doing this would be to calculate the yearly change from the transient model output, downscale these incremental changes to the region of interest and apply to the baseline climate data. This time consuming approach has not yet been applied in practice. Another approach is to calculate changes averaged over a period, such as a decade (e.g. calculate average change over the years 65 to 74, IPCC, 1992). The problem here is that there is considerable year-to-year variability in simulated (and observed) climate and that an average over a short period may not be representative of an underlying trend. A further major complication with transient simulations is 'model drift'. Over time, climate simulation models 'drift', producing for example a gradual increase in temperature (of a few tenths of a degree) over the duration of the simulation, even with stable boundary conditions. This drift will be present in a transient simulation and will either exaggerate or dampen any trend. The effects of model drift can be removed by subtracting a trend equal to the drift as determined from a control run simulation of the same length, or can be minimised by comparing the same periods in the control and transient runs; this means that two short periods are being compared and the estimated change will be strongly affected by sampling uncertainties. A final complication is known as the 'cold-start' problem. For the first few years of a transient simulation, there tends to be very little change in climate. This is attributed to problems in initialising the climate model. In summary, whilst transient GCM experiments appear to provide useful information on the rate and pattern of change over time, it is currently difficult to define credible scenarios for impact assessments from transient simulations.

Equilibrium GCM experiments simulate climate with CO_2 concentrations stabilised at twice current values. No dates can be assigned to such simulations, partly because CO_2 concentrations will not stabilise at a given level, and partly because the rate at which CO_2 concentrations increase is determined by emissions scenarios. Transient experiments simulate change over time, but again it is not feasible to assign directly dates to output. The 'cold start' problem causes a delay and the rate of change of simulated climate is a function of the assumed rate of change of greenhouse gas concentrations (usually, but not always, 1% compound per year). It is possible to develop scenarios for specified time periods from equilibrium or transient GCM output using a relatively simple global one-dimensional energy balance model which can determine global mean temperature over time under a given emissions scenario. Such a model — STUGE — is described by Wigley and Raper (1992) and Figure 3.11 shows the transient increase in global temperature simulated by STUGE under the six IPCC 1992 emissions scenarios, for an equilibrium climate change of 2.5°C.

Scenarios for specific dates can be derived from equilibrium GCM output as follows (Viner and Hulme, 1993):

1 Divide the change in climate (temperature, precipitation or any other measure) as simulated by an equilibrium GCM by the climate sensitivity of that GCM: the climate sensitivity is the change in global mean annual temperature under stable $2 \times CO_2$ conditions. This produces "change per degree of global warming".
2 Determine the change in global mean annual temperature for the year of interest, under the given emissions scenario.
3 Rescale the standardised GCM changes by the global mean annual temperature change, to determine the regional climate change by that date.

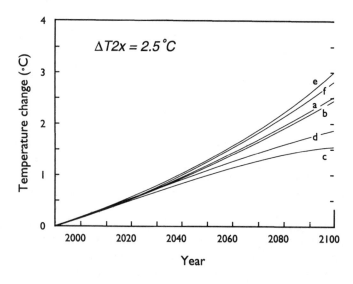

FIGURE 3.11 Change in global mean temperature, using STUGE, under the six IPCC 1992 emissions scenarios (Wigley and Raper, 1992).

This approach can be used to define "stable" scenarios, representing mean conditions at a given point in time, or transient scenarios defining change over time. It is simple, but makes the fundamental assumptions that the pattern of change is consistent over time and that the change in each climate parameter is a linear function of change in global mean annual temperature. Neither of these assumptions are likely to be true in practice and transient GCM experiments have shown that the pattern of change does vary over time as oceans respond slowly to altered conditions.

Time-specific scenarios can be derived from transient GCM output by selecting a period where the transient GCM change in global mean temperature is equal to the change as calculated using the energy balance model. Different periods will be selected for different transient GCM simulations. The approach assumes that the difference in rate of change between that simulated by the energy balance model and that assumed in the transient experiment does not significantly affect the pattern of change.

The primary problem with using GCM output to define climate change scenarios for hydrological impact studies is the coarse spatial resolution of current GCMs. As outlined above there are several methods for downscaling to the catchment scale. However, the derived catchment changes are, irrespective of method, fundamentally affected by the quality of the coarse scale climate simulation. The downscaling method, however sophisticated, may simply be adding spurious fine resolution detail to poor quality input information.

Hydrological studies require information on both daily climatic variability and changes in that variability and this cannot be taken directly from GCM output. It is possible to define scenarios which combine information on changes in monthly totals with arbitrary assumptions about changes in day-to-day frequencies, but three of the downscaling methods outlined above have the potential for creating credible daily-scale scenarios. Karl *et al.*'s (1990) analogue of the weather forecasting approach has potential, but would be cumbersome to apply over a large area: it requires large amounts of local data and calibration. Approaches based on circulation types are being used in hydrological impact studies (e.g. Bardossy and Caspary, 1991; Wilby *et al.*, 1993), but it may be difficult to define stable relationships which apply under both current and changed climatic conditions. Nested regional climate models appear to offer the best prospects for development of regional climate change scenarios at scales relevant to hydrologists (Hostetler and Giorgi, 1993). They provide spatial information and are not based on calibration. They do, of course, rely on the quality of simulation of large-scale climatic features.

The majority of climate change impact assessments have used equilibrium scenarios, representing mean conditions at some point in the future (usually when CO_2 is doubled, sometimes at a defined period). Although a future with increasing greenhouse gas concentrations will not be stable and a long-term mean is not particularly realistic, such scenarios do provide information about changes in risk or expected system properties in the future. Transient scenarios, incorporating a gradual change over time, give information about the rate of change and when important thresholds might be crossed, but simulated changes in risk or system reliability may be very difficult to interpret because of the underlying trend and the large year-to-year variability. It would be better to estimate changes in the risk of drought during the 2030s, for example, using a 50-year stable time series representing mean 2030s conditions (which would of course only last for a decade), than using the simulated transient flows over the 10 years between 2031 and 2040.

Change in potential evaporation

Climate change scenarios tend to be expressed in terms of changes in temperature and precipitation. However, hydrological impact assessments really need information on changes in evaporation and precipitation.

Evaporation is the transformation of water from liquid to vapour, and transpiration is the part of total evaporation that enters the atmosphere through plants (Shuttleworth, 1993); evaporation from a vegetated surface is often termed evapotranspiraton. Potential evaporation is the evaporation that would occur from an extensive surface (conventionally defined to be grass-covered), with no shortage of available water. Actual evaporation falls below the potential rate when there is insufficient water. Potential evaporation is a function of the energy available for evaporation, the rate at which water vapour can be removed from the evaporating surface, and plant physiological characteristics. All these factors may be affected by climate change.

The maximum energy available for evaporation will not be affected by climate change, because it is determined by the position and passage of the sun, except insofar as there might be an increase in the advection of energy. However, net radiation, which drives evaporation, will change. Cloud cover affects albedo and the proportion of total short-wave incoming radiation that reaches the land surface. Downward long-wave radiation is affected by atmospheric properties, including humidity and atmospheric CO_2 content (the greenhouse effect), and outgoing long-wave radiation emitted by the land surface is affected by surface temperature. Bultot *et al.* (1988b) derived a series of differential equations to predict changes in the parameters of an empirical model for estimating net terrestrial radiation in Belgium from changes in temperature, absolute humidity, net solar radiation and cloud cover. With an increase in temperature, a decrease in absolute humidity and an increase in cloud cover, they computed a 7% reduction in net terrestrial radiation and hence an increase in the amount of energy available for evaporation. In contrast, Lockwood (1993) simulated an increase in the long-wave radiation emitted from a warmer grass surface and a slight decrease in the amount of energy available for evaporation.

The rate at which water vapour can be removed from an evaporating surface is a function of the humidity of the air, which determines the rate at which vapour can be accepted by the air, and wind speed, which controls the rate of removal of air across the evaporating surface. Plant properties, such as size and leaf area index, affect turbulence in and above the plant canopy, and hence the rate of removal of water vapour. As air temperature increases, air can hold a greater amount of water vapour (between 5% and 6% more per degree Celsius): relative humidity therefore declines and evaporation can, unless limited by available energy, increase.

Plant physiological properties affect not only aerodynamic roughness, but also the rate at which water evaporates through plant transpiration. Plants lose water through stomata. Stomatal conductance rates (or their inverse, stomatal resistance rates) are affected by plant type and growth stage, water availability and, for many plants, atmospheric characteristics. Coniferous forest stomatal conductance rates increase as temperature increases and relative humidity declines, (e.g. Lockwood, 1993). Changes in temperature and precipitation will affect plant growth through the year, and therefore the seasonal variation in plant physiological properties. These changes may also lead to changes in the type of vegetation growing at a point or in a catchment.

Plant physiological properties are also affected by atmospheric CO_2 concentrations (Rosenberg *et al.*, 1990; Drake, 1992), in two main ways. First, increased CO_2 concentrations affect plant growth rates, depending on plant type. The C_3 group of plants (including most trees, legumes, root crops and small grains) use CO_2 relatively inefficiently and when CO_2 concentrations are increased photosynthetic fixation of carbon increases and plant growth is enhanced; a larger plant transpires a greater volume of water, so evaporation rates would increase. The C_4 group of plants (including most tropical grasses), however, use CO_2 more efficiently and are much less affected by increases in atmospheric concentrations. The second effect of CO_2 is on stomatal resistance. In general, when plants are exposed to higher CO_2 concentrations, their stomata close and transpiration is reduced (by perhaps a third: Kimball and Idso, 1983). These conclusions, however, are largely based on experiments made under controlled conditions and it is not yet clear whether they would be replicated under field conditions. Plant response to elevated CO_2 may be affected by climate or the availability of nutrients, and plants may adapt to altered CO_2 concentrations.

It is clear that the effects of global warming on potential evaporation are not simple. Many potential changes suggest an increase in potential evaporation, but these factors may be outweighed locally or regionally by factors reducing evaporation; the effects of higher temperature may be outweighed by an increase in humidity caused by a change in the passage of moist air masses; an increase in cloud cover might reduce net radiation enough to lower evaporation. The effects of given changes in all the factors affecting evaporation will also depend on local climatic characteristics. A specified increase in humidity will have little effect in an arid environment where humidity does not constrain evaporation, but will have a major effect on evaporation in wetter areas where the ability of air to accept moisture is a constraint.

It is difficult and expensive to measure potential evaporation, so many calculation schemes have been devised (Shuttleworth, 1993). The Penman formula (Penman, 1948) is theoretically based and combines energy balance and aerodynamic processes. It requires data on net radiation, temperature, humidity and wind speed, and is generally accepted as the standard method for estimating potential evaporation. The Penman-Monteith (Monteith, 1965) formula is an extension which incorporates the effect of plant physiological properties. Priestley and Taylor (1972) presented a simplified form of the Penman equation, based on the observation that the aerodynamic component of evaporation was generally a fixed proportion of the energy balance component; their equation uses just net radiation and temperature. Other equations are much more empirical. The Thornthwaite (1948) formula uses just temperature and day length, as does the Blaney-Criddle (Blaney and Criddle, 1950) equation in its original form.

Martin *et al.* (1989) used the physically-based Penman-Monteith formula to investigate the relative importance of changes in each controlling variable on potential evaporation, for wheat, grassland and forest sites in Nebraska, Kansas and Tennessee respectively. Figure 3.12 shows the change in grassland potential evaporation, on a summer day, for a given change in temperature, net radiation and humidity. Potential evaporation was found to be most sensitive to changes in temperature, net radiation and stomatal resistance, and least sensitive to wind speed, for all three sites, although the degree of sensitivity varied between the sites and from day to day.

The estimated effect of a change in climate on potential evaporation depends on the characteristics of the site, the equation used and, if the equation uses more than one climatic input, the relative changes in all the factors affecting evaporation. Different hydrological impact studies have used different equations and assumptions. Nemec and Schaake (1982) and Nash and Gleick (1991), for example, both used Budyko's (1982) estimate of 4% increase in potential evaporation per degree Celsius, derived from radiation balance calculations. Cohen (1987a; 1991) and Gleick (1987) both used the Thornthwaite formula, whilst Wolock *et al.* (1993) used a variant on the Thornthwaite formula. Mimikou *et al.* (1991) and Palutikof *et al.*

(1994) used the Blaney-Criddle formula. Lettenmaier and Gan (1990) calculated potential evaporation using the Penman formula—assuming no change in factors other than temperature — whilst Bates *et al.* (1994) used the Priestley-Taylor simplification of the Penman formula and assumed changes in both temperature and net radiation. Bultot *et al.* (1988a; 1992) used an energy balance approach to estimate changes in potential evaporation in three Belgian and one Swiss catchment. Their estimated changes are shown in Table 3.2. They assumed changes in radiation components and temperature, but no change in relative humidity.

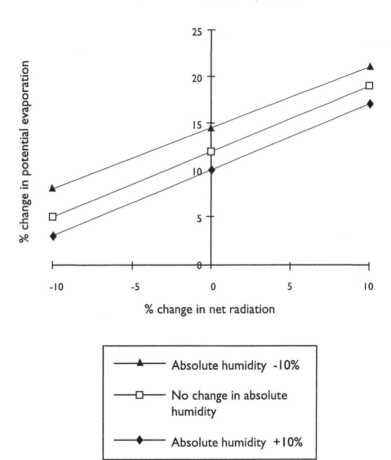

Grassland: Konza Prairie, Kansas

FIGURE 3.12 Sensitivity of summer potential evaporation from grassland to change in temperature, net radiation and absolute humidity (data from Martin *et al.*, 1989).

The estimated sensitivity of potential evaporation to change in temperature depends very much on the formula used. Both the Thornthwaite and the Blaney-Criddle formulae tend to produce a higher change in potential evaporation for a given change in temperature than the Penman formula. Table 3.3a shows percentage change in potential evaporation in southern England for a 4°C increase in temperature as determined using the Penman, Thornthwaite and Blaney-Criddle formulae (Arnell *et al.*, 1990), and Table 3.3b shows estimated change in annual potential evaporation in four catchments using four estimation equations (Yates and Strzepek, 1994a).

	J	F	M	A	M	J	J	A	S	O	N	D	Ann.
Change in temperature (°C)													
	3.1	3.4	3.4	3.1	2.8	2.7	2.5	2.3	2.3	2.7	2.8	3.2	2.9
Change in potential evaporation (%)													
Zwalm[1]	19	30	18	10	6	8	6	6	2	6	17	21	9
Dyle[1]	22	29	19	10	6	8	7	6	3	7	19	22	9
Semois[1]	33	45	24	11	7	8	7	7	3	8	25	23	9
Murg[2]	36	56	33	12	7	8	6	6	4	9	25	19	10

notes: [1]Belgium: Bultot *et al.* (1988a) [2]Switzerland: Bultot *et al.* (1992)

TABLE 3.2 Change in potential evaporation in Belgian and Swiss catchments (Bultot *et al.*, 1988a; 1992)

a)	J	F	M	A	M	J	J	A	S	O	N	D
Penman	23	17	15	13	11	9	9	9	10	13	18	21
Thornthwaite	50	45	26	18	15	15	16	16	15	15	22	37
Blaney-Criddle	61	60	52	44	37	32	30	31	33	38	49	57

b)	Blue Nile, Ethiopia	Vistula Poland	Mulberry Arkansas	East Colorado
Penman	8	11	10	32
Priestley-Taylor	8	10	8	29
Thornthwaite	49	22	36	24
Blaney-Criddle	25	17	13	23

TABLE 3.3 Effect of change in temperature on potential evaporation: different estimation equations a) Southern England (Arnell *et al.*, 1990): 4°C increase in temperature. Percentage change in monthly potential evaporation; b) Four catchments (Yates and Strzepek, 1994a): 4°C increase in temperature. Percentage change in annual potential evaporation

The difference in estimated change in potential evaporation makes it difficult to compare the results of studies using the same scenario for change in temperature, but different estimation procedures.

MODELS FOR IMPACT ASSESSMENTS

Hydrological models

A hydrological model translates climatic inputs into hydrological outputs. Models can be classified according to purpose (for example forecasting, design or long-term simulation) or characteristics, and the most convenient classification distinguishes between empirical, conceptual and physics-based models.

Empirical models relate climate inputs to hydrological properties through an empirical relationship, such as one based on regression analysis, without prescribing the physical processes involved. Empirical relationships can be derived to predict long-term average hydrological characteristics (most usually average annual runoff, but also characteristics such as the mean annual flood or the mean annual minimum discharge) or to forecast flows from preceding flow, rainfall and temperature (as in some operational river flow forecast models).

Langbein (1949) derived relationships between annual rainfall, annual temperature and annual runoff in the United States, and these relationships were used by Revelle and Waggoner (1983) to estimate the effects on annual runoff of changes in rainfall and temperature: such empirical models are in a sense spatial analogues. However, the implied sensitivity of runoff to changes in inputs depends significantly on the form of the empirical model and the model parameters (which may depend on the period of record used). Arnell (1992), for example, found very significant differences in estimated changes in average annual runoff in the UK using different regression and other empirical models predicting runoff from rainfall and potential evaporation. Empirical models internalise relationships between the independent variables which together control the dependent variable, and these internal relationships (such as that implied between potential and actual evaporation) may not apply under altered climatic characteristics. Also, the effect of a change in a controlling variable (such as annual rainfall) may depend on how that change is distributed over time; in the UK, a given percentage change in annual rainfall would have a greater effect, the heavier the concentration of the increase in winter (Arnell, 1992). Leavesley (1994) notes that "extension of these relations to climate or basin conditions different from those used for development of the function is questionable" (p161).

Many flow forecasting models are based on empirical relationships between flow at one time period and flow, rainfall and temperature in earlier periods. Wright (1980), for example, developed a series of empirical regression models to estimate monthly flow in several British rivers from flow, rainfall and evaporation in previous months. Arnell and Reynard (1989) subsequently used Wright's model to simulate the effects of changes in rainfall and evaporation on monthly river flows, but the simulated changes were very much dependent on the parameters of the catchment model and the implied internal relationships between variables and months.

Conceptual models are based on an idealised representation of a catchment as a series of stores, and work through an accounting process to determine the water balance at different time scales. Parameters define movement of water into and out of these stores and although these parameters might have notional physical interpretations, in practice they must generally be calibrated using observed river flow data. Conceptual models have been very widely used in studies estimating the impacts of climate and other environmental changes on river flows. Some studies have used a monthly time step (Cohen, 1987a, 1991; Gleick, 1987; Mimikou et al., 1991; Arnell et al., 1990; Kwadijk, 1993), whilst others have worked on a daily or shorter time step (Nemec and Schaake, 1982; Lettenmaier and Gan, 1990; Bultot et al., 1988a; 1992; Nash and Gleick, 1991; Vehvilainen and Lohvansuu, 1991; Kite, 1993b; Bates et al., 1994). The choice of time step is essentially a function of catchment size and data availability. Hydrological processes operate on a time scale of minutes — during rainfall events — but as the size of catchment increases, the fine temporal details become less and less important: spatial variability increases, and the catchment itself smooths out short-term temporal variations. Studies in large catchments (over 10 000 km^2), or where daily climatic data are not available, tend to use a monthly time step; studies in smaller catchments use a daily time step. Lumped conceptual models treat the catchment as a single unit, whilst distributed conceptual models break a catchment into sub-catchments.

The third category includes models which use physics-based equations to simulate the movement of water into and through the soil, and along the hillslope into the river. An example is the SHE (Système Hydrologique Européen) model (Abbott et al., 1986), which operates at a time resolution of minutes and a spatial scale down to metres. In principle, the parameters of such physics-based models can be determined by measurement of physical properties (such as soil hydraulic conductivity), but in practice many of the parameters must be based on calibration. Such models have not been used in climate change impact assessments, although a simple physics-based model — TOPMODEL — was used by Wolock and Hornberger (1991) to simulate flows in an Appalachian catchment under changed climates. The main reasons why physics-based models have not been used to estimate the effects of climate change are that they are very data intensive, and that it would be necessary to define climate change scenarios at very fine temporal scales: this is not feasible at present. Studies into the potential effects of changes in catchment physical properties — such as soil structure — on hydrological behaviour would need to use physics-based models.

Most climate change impact studies have some form of conceptual hydrological model, operating at either a daily or monthly time step, because physics-based models are too complicated and empirical models too unreliable for estimating effects of change. Yates and Strzepek (1994b) evaluated the suitability of two conceptual water balance models, two empirical models estimating annual runoff and one empirical model predicting monthly runoff from runoff in previous months, using five study catchments. They concluded that the lumped conceptual models were more appropriate for climate change impact assessments than the empirical models, for two main reasons. First, the conceptual models were much more robust, and were able to simulate flows in the validation period nearly as well as during the calibration period; the performance of the monthly empirical model was considerably worse during the validation period, emphasising how the model parameters were heavily dependent on the period of record used for calibration. Second, the two conceptual models gave much more consistent estimates of sensitivity to change in climatic inputs than the regression models. Figure 3.13 shows the change in average annual runoff for an increase in temperature of 2°C and different changes in annual precipitation, in two catchments, with different water balance and empirical models.

The key assumption when using models with calibrated parameters in climate change assessments is that the model, or more specifically model parameters, will remain valid under changed climatic conditions. The robustness of model parameters, and the robustness of model performance, can be determined by estimating model parameters from different periods of record — split-sample testing — or by examining the performance of the model during anomalous climatic periods — differential split-sample testing (Klemes, 1986; Gleick, 1986; Leavesley, 1994). Model validation is an essential component of a climate change impact assessment. Credible assessments can only be based on credible hydrological models, which have been shown both to reproduce observed hydrological characteristics and to represent hydrological processes in a realistic manner.

Water resource models for impact assessment

Hydrological flow and recharge models simulate only the effects of climate change on hydrological regimes; the consequences of these effects for a system (reservoir, water user or aquatic ecosystem, for example) must be assessed using additional models.

Water supply and distribution models are widely used by water managers and can, in principle, be used to simulate the consequences of climate change. These models range from simple single-reservoir operational models, where the output is simulated yield and risk of reservoir failure, to complicated

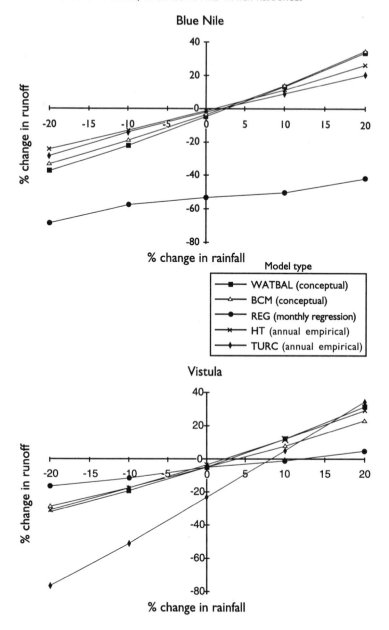

FIGURE 3.13 Effect of model on estimated sensitivity of average annual runoff to change in temperature and precipitation, for two catchments (Yates and Strzepek, 1994b)

multi-source conjunctive use models, where the mix of sources exploited at any one time — and hence the risk of system "failure" — is dependent on factors such as water availability, the distribution of demands over space, the relative operating cost of different sources, and the characteristics of the interconnected system. Water system simulation models are site-specific, but the most robust have a form similar in principle to that of a conceptual water balance model.

Models for other water users and other aspects of the water system are much less well defined at present. Mechanistic ecological models are being developed which translate changes in system inputs into ecological consequences (Malanson, 1993), but many ecological models currently used in climate change impact assessments are based on empirical statistical relationships (Parr and Eatherall, 1994). The problems involved in estimating the consequences of changes in hydrological regimes are returned to in Chapter 8.

IMPACTS AND ADAPTATIONS

Biophysical and socio-economic effects and impacts

Climate change has a physical *effect* on a system. The *impact* of this effect depends on the value attached to the change. A large physical effect may have a small impact: conversely, a small physical effect may have a very large impact.

Hydrological studies tend to express effects in terms of either volumetric or percentage change in runoff volumes, over time scales ranging from daily, through weekly to monthly, seasonal and annual. Some (e.g. Bultot *et al.* (1992), Switzerland) evaluated changes in the magnitude of events with a specific frequency, while others (Gellens (1991), Belgium) have determined changes in the rate at which certain discharge thresholds are passed.

Studies focusing on water resource systems either examine changes in the quantity (of water or power) supplied, or in the risk of meeting current design and operational targets. Many of these studies have made qualitative assessments of the economic and social impacts, but very few have attempted quantitative assessments. Examples include Haneman and McCann (1993) who estimated the cost of lost hydropower production in north California under one change scenario to be $145M per year (because the lost power was made up by burning more expensive gas), and Kirshen and Fennessey (1993) who estimated that the extra sources needed to maintain supplies to Boston, Massachusetts, under one warming scenario would cost around $722M to implement and another $21M a year to run. There are several major conceptual problems in estimating the full economic costs of a change in water resources (as noted in both the studies cited above). The first is that the costs of change depend on the management response to change (Arnell and Dubourg, 1995). In the most general terms, there are three different responses. At one extreme, there is no direct adaptation to climate change. This might mean that some uses of water become impossible, and the economic costs are therefore the lost value of that water to those users. It might be the case that other uses become possible — perhaps because the amount of water available increases — and in these circumstances the economic benefit of change will equal the value of the new activities. At the other extreme, the water management system may adapt to ensure that defined targets are still met; the costs of change are therefore the costs of continuing to meet these defined targets (in the Boston case study, the costs of continuing to supply water to Boston with a given reliability). Of course, climate change might lower the cost of meeting current standards. The third management strategy is between these two extremes. It recognises that some response to change would be economically justifiable, but that attempting to maintain current standards would not be; the appropriate level of adaptation would be that which is economically most efficient.

A second problem lies in the distinction between financial costs to an organisation and economic costs to a specific geographical region. Some organisations may gain whilst others loses; similarly costs in one region may be offset by benefits in another.

A third conceptual problem arises because climate change impacts occur in the future (possibly several decades away), and need to be expressed in terms of present prices and values. The discount rate defines the preference of the economic decision-maker for income now rather in the future. The higher the discount rate, the greater the weight given to short-term income; the lower the rate, the greater the preference for income over the longer term. Considerable attention is being paid by economists to the definition of appropriate discount rates for climate change impact studies.

Autonomous adjustments

"Autonomous adjustments" are defined to be alterations to management practices that occur for reasons other than climate change. These changes will affect the management system, and may well affect its sensitivity to change. Increasing flexibility in water management due to increased inter-linkage of different water supply sources, for example, might be expected to improve the ability of the water supply system to withstand shortages and reduce sensitivity to change. On the other hand, increasingly rigid and strict environmental standards may make a water management authority more sensitive to climate change; previously any deterioration would not have mattered.

Ideally, a climate change impact assessment would allow for the effects of such autonomous management adjustments, but in practice this is very difficult. Many adjustments are operational responses to unusual conditions and may be very difficult to predict. The water industry in Britain responded very differently to the drought of 1988 to 1992 than to the 1976 drought, partly because many of the lessons of the earlier drought had been learnt. Other adjustments reflect longer-term policy changes, and these may be defined by the water management scenarios developed under Step 4.

Figure 3.14 summarises the effect of various adaptations and human interactions on how a climate change is translated into an impact on a water management agency (Arnell, 1995a).

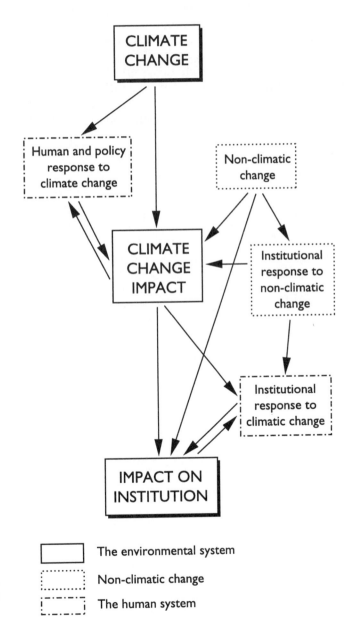

Figure 3.14 Impact of global warming on a water management agency (Arnell, 1995a)

Adaptation strategies

The final step in a comprehensive impact assessment is to evaluate different strategies for adapting to the changes caused by global warming. In principle, it is possible to compare the costs or environmental impacts of different responses to a given change scenario, and to both define feasible strategies and determine when such strategies would need to be implemented. In practice, this is very difficult because there are frequently many different feasible strategies, and of course there may be several different change scenarios. A study would be restricted to a broad assessment of alternatives (would it be necessary to build a new reservoir, for example, or could any possible change be accommodated by changing operating rules of the current system?), and would need to consider explicitly the uncertainty introduced by the use of several change scenarios. In fact, the robustness of a given adaptation strategy to different possible futures could be a means of selecting an appropriate strategy. Possible adaptation strategies will depend on management goals, specific criteria and institutional, social and practical constraints.

CHAPTER 4

The Great Britain case study: catchments, models and scenarios

The Great Britain case study investigates the impacts of future climate change on river flows. This chapter introduces the case study in terms of the steps of an impact assessment outlined in Chapter 3, before summarising current hydrological regimes in Britain. The study catchments are described, and the hydrological model and the climate change scenarios defined.

METHODOLOGY

Chapter 3 introduced the seven steps in a climate change impact assessment (Carter *et al.*, 1994). The problem addressed in the study (Step 1) was to determine the potential effects of climate change on river flow regimes in British catchments. The study was to consider catchments across the whole of Great Britain (England, Scotland and Wales) covering the full range of hydrological regimes. The catchments were to be unaffected by major water management activities such as reservoir impoundment, and to have minimal abstraction of water or discharge of sewage effluent. The study did not consider the effects of land use change on hydrological regimes.

The method selected (Step 2) was to apply a catchment hydrological model with climatic inputs (precipitation, temperature and potential evaporation) perturbed according to scenarios derived from climate simulation models (see the bottom right-hand side of Table 3.1). This approach was adopted primarily to ensure that estimated *hydrological* changes were consistent with the *climatic* changes estimated for a given period by climate models. Climate model runoff simulations were not used directly because climate model grid cells are very large relative to British catchments. The temporal analogue approach was not used because different analogues give very different results and past climates are not necessarily good analogues for future climate (Arnell *et al.*, 1990); neither were spatial analogues considered, because hydrological regimes are a function of both regional and local influences, and it is not possible simply to transfer hydrological information from one region to another (Arnell *et al.*, 1990). Empirical relationships between hydrological characteristics and catchment and climate properties were not used to estimate the effects of climate change, because the estimated effects are dependent on the form and parameterisation of the empirical model.

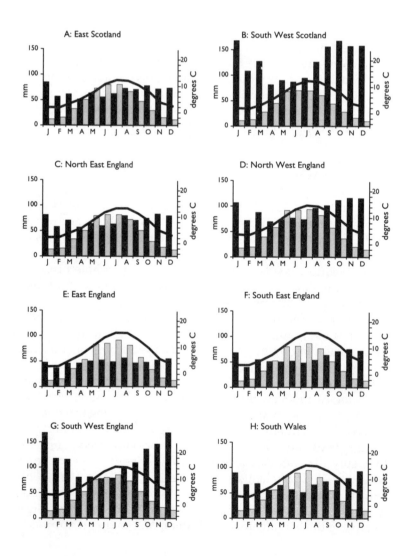

FIGURE 4.1 Monthly mean precipitation (dark bar), temperature (line) and potential evaporation (light bar) for eight locations in Great Britain

The hydrological model used to simulate river flows from precipitation, temperature and potential evaporation data is described below. A major part of the study was the development of the hydrological model and ensuring that it simulated river flows accurately in the study catchments (Step 3).

The study took the period 1961 to 1990 as a baseline, and used the climate change scenarios developed for the 1996 Climate Change Impacts Review Group (CCIRG, 1996; Hulme, 1996). These scenarios, outlined below, were based on output from the UK Hadley Centre transient climate change experiment (Murphy, 1995; Murphy and Mitchell, 1995). Both equilibrium scenarios, representing conditions in the 2050s, and transient scenarios covering the period 1990 to 2050 were used. It was assumed that all other catchment properties, including land use, soil properties and human intervention, remained constant over time.

The case study focused on the hydrological effects of climate change (Step 5), and these effects were expressed in terms of changes in the volume of river runoff (over a month, season and year), the magnitude of discharges occurring with a given frequency, and the frequency of occurrence of discharges of a given magnitude. The *impacts* of these changes on water uses and water resource management were not assessed directly, but implications for water resources were drawn, particularly in the context of other changes affecting water over the next few decades (Step 6). These, together with some discussion of adaptation (Step 7), are explored in detail in Chapter 8.

CLIMATE AND CONTROLS ON HYDROLOGICAL REGIMES IN BRITAIN

Great Britain has a humid temperate climate. Average annual rainfall ranges from 600 mm per year in the east to over 2500 mm in some upland areas in the north and west. Rainfall is generally evenly distributed through the year. Annual potential evaporation varies between 450 mm and 620 mm, with the highest values in the south and east. Average annual rainfall exceeds average annual potential evaporation across all of Great Britain (by as little as 15 mm in parts of the east), although in some dry, warm years annual rainfall is less than annual potential evaporation. Potential evaporation exceeds rainfall during summer, so most river flow regimes show a peak in winter. Across all but upland Britain, very little precipitation falls as snow. Figure 4.1 shows monthly average rainfall, temperature and potential evaporation, again over the period 1961—1990, for six locations across Britain (Figure 4.2. Data from the Meteorological Office MORECS data base Thompson *et al.*, 1981).

Catchment physical properties such as geology and soil characteristics have a major influence on river flow regimes in Britain, and are largely responsible for the considerable variability in flow regime between catchments. Figure 4.3 shows the distribution of the three major aquifers in Britain. All are recharged primarily during winter and all sustain river flows through summer, but have different river flow regimes. There is very little direct runoff from catchments underlain by chalk, and virtually all the effective precipitation reaches the groundwater store, from which it drains slowly. There is therefore relatively little variation in flow through the year. In some chalk aquifers, it may take several years for particularly high or low winter recharge to affect river flows. Limestone aquifers, however, may be recharged very quickly through fissures, and river flows respond more quickly to rainfall.

FIGURE 4.2 Locations of example climate stations

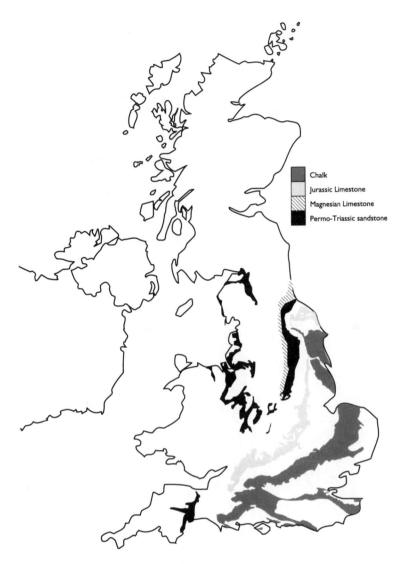

Chalk
Jurassic Limestone
Magnesian Limestone
Permo-Triassic sandstone

FIGURE 4.3 The chalk, limestone and Permo-Triassic sandstone aquifers in Britain

THE STUDY CATCHMENTS

Twenty-one catchments were selected for analysis. They were chosen using four criteria:

- The river flow data had to be of high quality, particularly at low flows: the assessment of gauging station quality used the classification developed by Gustard *et al.* (1992);

- The flow records had to be relatively unaffected by human influences, such as water abstraction and effluent discharge: this assessment was also based on Gustard *et al.* (1992);

- The flow records had to span at least the period 1980 to 1989: data from this period were used for model calibration and validation;

- The total sample of catchments had to represent all of the different flow regime and catchment types found in England, Scotland and Wales.

The selected catchments are plotted in Figure 4.4 and summarised in Table 4.1. The Base Flow Index (BFI: Institute of Hydrology, 1980) is calculated from the daily flow data, and is assumed to index the relative contributions of "baseflow" and

FIGURE 4.4 Study catchments

"direct runoff" to total streamflow. The higher the BFI, the greater the contribution of baseflow to streamflow, and the less the day-to-day variability in flow. Catchments underlain with significant groundwater aquifers tend to have high BFIs, with the highest values (>0.9) characteristic of catchments underlain by chalk.

Table 4.1 also shows average annual precipitation, potential evaporation and runoff, over the period 1980 to 1989, together with the runoff coefficient, which is the ratio of average annual runoff to average annual rainfall: the lower the coefficient, the more "arid" the catchment. The table also indicates those catchments in which an important proportion of precipitation falls as snow. Figure 4.5 shows daily flows for 1989 for six of the study catchments. Maximum flows tend to occur during winter and spring, and minima occur in late summer. The catchments with the most impermeable bedrock show the greatest variation from day to day (and have the lowest BFI), while the chalk catchments, such as the Lambourn, on permeable geology have very little variability.

TABLE 4.1   The study catchments

National Water Archive code	River and gauging station	Basin area (km²)	Base Flow Index (BFI)	Average annual 1980 -1989 rainfall (mm)	potential evaporation (mm)	runoff (mm)	Runoff coefficient
11001	Don at Parkhill*	1273.0	0.67	889	478	535	0.60
19001	Almond at Craigie Hall*	369.0	0.38	968	482	564	0.58
21018	Lyne Water at Lyne Station*	175.0	0.59	1019	482	571	0.56
24004	Bedburn Beck at Bedburn*	74.9	0.46	871	507	536	0.61
25006	Greta at Rutherford Bridge*	86.1	0.21	1271	521	862	0.68
28008	Dove at Rocester Weir	399.0	0.61	1042	558	650	0.62
32003	Harpers Brook at Old Mill Bridge	74.3	0.49	645	565	189	0.29
34004	Wensum at Costessey Mill	536.1	0.73	692	582	268	0.39
37005	Colne at Lexden	238.2	0.53	617	601	151	0.24
38021	Turkey Brook at Albany Park	42.2	0.21	691	583	163	0.24
39008	Thames at Eynsham	1616.2	0.68	756	581	274	0.36
39019	Lambourn at Shaw	234.1	0.96	755	584	238	0.31
40007	Medway at Chafford Weir	255.1	0.50	840	569	380	0.45
43005	Wiltshire Avon at Amesbury	323.7	0.91	764	599	338	0.44
47001	Tamar at Gunnislake	916.9	0.46	1277	588	772	0.60
54008	Teme at Tenbury	1134.4	0.57	894	559	423	0.47
57004	Cynon at Abercynon*	106.0	0.42	1882	541	1407	0.75
58009	Ewenny at Keepers Lodge*	62.5	0.58	1439	596	1056	0.73
66011	Conwy at Cwm Llanerch*	344.5	0.29	2204	518	1786	0.81
76005	Eden at Temple Sowerby*	616.4	0.37	1253	496	754	0.60
79006	Nith at Drumlanrig*	471.0	0.34	1684	462	1201	0.71

* Affected by snow

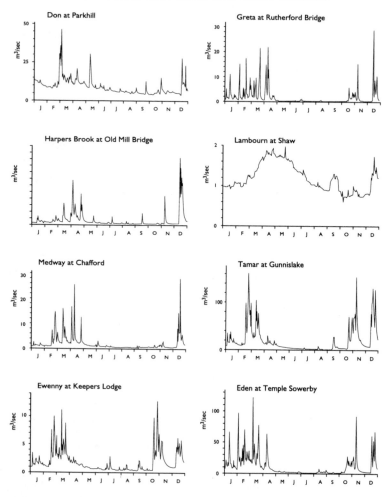

FIGURE 4.5 Daily flows for eight of the study catchments in 1990

RIVER FLOW SIMULATION

A lumped conceptual daily rainfall-runoff model simulated the translation of rainfall and potential evaporation into streamflow, incorporating in some catchments the effect of snowfall and snowmelt. This section outlines the rainfall-runoff and snow models used, and describes the results of a split-sample model validation exercise. The model was applied in the 21 study catchments and also, in a generalised form, across the whole of Great Britain, treating 0.5° × 0.5° grid cells as catchments.

The rainfall-runoff model

The requirement to apply the model to many catchments excluded the most complex physics-based models from consideration, while the need to estimate the implications of changes in inputs eliminated empirical models. A range of lumped conceptual models was therefore explored: each model had parameters representing some physical process calibrated from observed flow and climate data.

Each model consisted of a number of stores, with parameters controlling store size and the rate of removal of water from the store, and operated at a daily time step. The fundamental model equation is

$$Y_i = P_i - E_i + \Delta S$$

where Y_i is runoff in time step i, P_i is precipitation, E_i is evaporation and ΔS is the change in storage as a result of infiltration and drainage. There are four basic components to a conceptual rainfall-runoff model:

- A procedure to determine actual evaporation from potential evaporation. The ratio of actual to potential evaporation is generally taken to be a function of the contents of one of the soil stores. Some models use a linear function (with a linear decline as soil moisture content falls below some maximum capacity), whilst others use an negative-exponential function which means that the ratio of actual to potential evaporation falls slowly at first, but falls more rapidly as the soil store becomes more and more empty.

- A runoff generation procedure, converting excess precipitation into runoff. Most models assume that any water in the soil in excess of field capacity becomes runoff (often termed direct runoff), and some also allow for drainage out of the bottom of the soil into the river via groundwater. A few models also generate runoff when rainfall intensity exceeds some infiltration capacity.

- A storage accounting procedure, determining the contents of each storage. Store contents at the end of any time step are based on the contents at the beginning of the step and on inflows and outflows during the time step. Different models have different numbers of stores (the outflow from one is usually the inflow into another) and different procedures for determining outflows (which are usually dependent on store contents). Some stores can overflow; others can only drain downwards (and actual evaporation is of course an outflow from a storage).

- A routing procedure, translating the outputs from each store into the river channel.

Moore's probability-distributed model (PDM: Moore, 1985) was selected after evaluation using a number of study catchments. The model assumes that soil moisture capacity varies across the basin (Figure 4.6) and therefore that the proportion of the basin with saturated soils varies over time. The model is similar in concept to the ARNO (Dümenil and Todini, 1992) and VIC (Wood *et al.*, 1992; Stamm *et al.*, 1994) models.

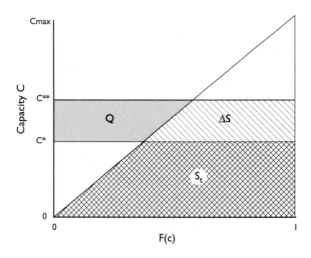

Figure 4.6   Variation in soil moisture capacity across the catchment

The version of the model used here assumes (like Moore, 1985) that the soil moisture capacity c is represented by the reflected power distribution:

$$F(c) = 1 - \left[\frac{1 - c}{c_{max}} \right]^{b} \quad 0 \leq c \leq c_{max}$$

capacity less than c, c_{max} is the maximum storage capacity in the catchment, and b defines the degree of spatial variability: 0 implies constant capacity, while 1 indicates uniform variation across the catchment (Figure 4.7). Soil moisture storage capacity is assumed to be saturation capacity.

The maximum amount of water that can be held in storage in the basin is the total area under the curve in Figure 4.6:

$$s_{max} = \int_{0}^{C_{max}} (1 - F(c))\, dc = \frac{c_{max}}{(1 + b)}$$

If all the stores with a capacity less than some critical value c* are full, then the volume of water stored in the catchment is

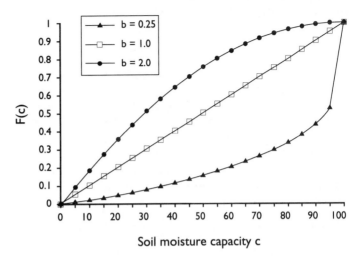

FIGURE 4.7 Effect of different values of the parameter b

indicated by the light-shaded area in Figure 4.6, and is equal to

$$S_i = \int_0^{c^*} (1 - F(c)) \, dc = S_{max}\left[1 - \left\{ \frac{1 - c^*}{c_{max}} \right\}^{b+1} \right]$$

For a given catchment total soil moisture storage s_i, the critical capacity c^*, below which all the stores are filled, is therefore

$$c^* = c_{max}\left[1 - \left[1 - \frac{s_i}{s_{max}} \right]^{\frac{1}{b+1}} \right]$$

Direct runoff is the volume of effective rainfall (rainfall minus evaporation) above that needed to saturate the soil, and essentially occurs from the portion of the catchment over which soils are filled to storage capacity. At the end of a day with rain, the critical capacity below which all stores are filled is

$$c^{*'} = c^* + P_i - E_i - Dr_i$$

where Dr_i is drainage from the soil. If $c^{*'}$ is less than c_{max}, then a portion of the effective rainfall goes to soil storage, whilst the portion falling on the "full" part of the catchment creates direct runoff Qd_i (Figure 4.6):

$$Qd_i = \int_{c^*}^{c^{*'}} F(c) \, dc = (P_i - E_i - Dr_i) - S_{max}\left[\left\{ 1 - \frac{c^*}{c_{max}} \right\}^{b+1} - \left\{ 1 - \frac{c^{*'}}{c_{max}} \right\}^{b+1} \right]$$

If $c^{*\prime}$ is greater than c_{max}, meaning that the entire catchment has reached capacity, then direct runoff is

$$Qd_i = (P_i - E_i - Dr_i) - (s_{max} - s_i)$$

The model assumes that the ratio of actual to potential evaporation declines as a negative-exponential function of basin average soil moisture deficit. Water drains out of the soil store into a groundwater store as a linear function of soil storage content. Baseflow occurs from the groundwater store, and direct runoff is routed through two cascading linear reservoirs to simulate channel flow. The model has five parameters, which are shown in Table 4.2.

The snow model

The hydrological regimes of catchments in upland and northern Britain are affected by snowfall and snowmelt. In some catchments these effects are important only in some years: in others they influence flow regimes in every year. The daily rainfall records for catchments in northern and western Britain (Table 4.1) were therefore adjusted to allow for precipitation falling as snow and subsequently melting.

TABLE 4.2 Parameters of the runoff generation model

Parameter	Description
c_{max}	Maximum soil moisture capacity in the basin (mm)
s_{max}	Maximum amount of soil moisture storage over the entire basin (mm): c_{max} and s_{max} together determine b, which describes the variability in soil moisture capacity across the basin
K_b	Soil drainage coefficient (mm/day)
G_{rout}	Baseflow routing coefficient
S_{rout}	Channel routing coefficient

The snow model used (Harding and Moore, 1988) assumes that precipitation falls as snow when temperature falls below some critical threshold, and that the snowpack begins to melt once another temperature threshold is passed. More particularly, it assumes that melting snow builds up within a snowpack and drains rapidly once the liquid water content of the snowpack exceeds some threshold. Rain falling onto the snowpack contributes directly to the liquid water content, and may therefore lead to a rapid increase in drainage from the snowpack. As the snowpack declines, an increasing proportion of the catchment becomes free of snow. Subsequent snowfall covers the entire catchment. Figure 4.8 (from Harding and Moore, 1988) summarises the model structure.

The total drainage from the snow store is therefore made up of "slow" and "fast" drainage, and is added to any rain falling on the snow-free part of the catchment to yield a volume of water available for runoff or replenishing soil moisture stores.

The model parameters are summarised in Table 4.3: they are the ones used in the operational flow forecasting system in Britain. The effect of the model is to adjust the input rainfall series to represent precipitation falling as rain and water released from a melting snow cover.

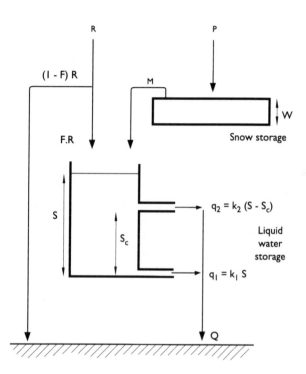

FIGURE 4.8 Structure of the snow-melt model (Harding and Moore, 1988)

TABLE 4.3 Parameters of the snow model

Parameter	Description	Value
T_{crit}	Threshold temperature for snowfall	1°C
T_{melt}	Threshold temperature for snowmelt	0°C
K_{melt}	Melt rate	4 mm °C^{-1} day^{-1}
S_c	Volume of snow pack at which entire catchment is snow-covered	100 mm
S^*	Proportion of liquid water in snow pack, above which drainage rate increases	0.04
K_1	Slow drainage rate	0.15
K_2	Fast drainage rate	0.85

Model calibration and validation

The model was calibrated over the period 1980 to 1983, and validated between 1983 and 1989. Daily river flow data were extracted from the National Water Archive's River Flow Archive (held at the Institute of Hydrology), and catchment average daily rainfall data were taken from the Institute of Hydrology's daily rainfall archive. Potential evaporation was determined from the monthly MORECS data set (Thompson *et al.*, 1981). MORECS (the Meteorological Office Rainfall and Evaporation Calculation System) uses the Penman-Monteith equation to estimate potential evaporation for a range of land covers: the values for a grass cover were used in the case study. Each catchment was allocated to one of the 40 × 40 km MORECS grid boxes, and the monthly box values were applied to the catchment without correction for differences in altitude between the catchment and the box. The monthly MORECS potential evaporation values were converted to daily values simply by dividing by the number of days in the month.

The snow model, applied to adjust the rainfall record in northern and upland catchments, required daily temperature data. Catchment temperature series were determined by correcting data from nearby temperature measurement sites for the difference in altitude using a standard lapse rate of 0.6°C per 100 m. The mean catchment altitude was calculated as the mean of the maximum altitude in the catchment and the altitude of the gauging station.

Parameter estimates and model evaluation

The five parameters of the PDM were estimated for the 21 study catchments from the calibration period spanning 1981 to 1983 (1980 was used as a wind-up period). The model evaluation was based on a comparison of observed and simulated flows during the calibration period (1981 to 1983) and a validation period (1984 to 1989: 1983 was used as the wind-up period).

First estimates of model parameters were determined by using an automatic Rosenbrock optimisation procedure to minimise the following objective function:

$$OJ = \frac{\sum_{i=1}^{N} (\log O_i - \log S_i)^2}{N}$$

where O_i is the observed flow on day i and S_i is the simulated flow. The logarithms of the flows were used to prevent the optimisation becoming biased towards the largest flows.

In the second stage some of the optimised parameters were manually adjusted to obtain a better visual fit, a closer match between observed and simulated annual runoff or to force the model to simulate a more realistic division between baseflow and direct runoff (for some groundwater catchments the optimised model parameters implied that baseflow provided only a small proportion of the total runoff).

The final model parameters are shown in Table 4.4, together with the Nash-Sutcliffe (1970) efficiency criterion calculated over both the calibration (1981 to 1983) and validation (1984 to 1989) periods:

$$Eff = 1.0 - \frac{\sum\limits_{i=1}^{N} (O_i - S_i)^2}{\sum\limits_{i=1}^{N} (O_i - \bar{O})^2}$$

where $\bar{O}$ is the observed mean flow. This criterion is biased towards large discharges, and it must be remembered that the optimisation procedure was not designed to maximise efficiency. The criterion is widely used, however, and gives an indication of model performance.

Model performance

The ability of the model to simulate river flow regimes was determined by comparing observed and simulated average annual runoff, monthly flow regimes, daily flow regimes for two example years (one in the calibration period, one in the validation period), and flow duration curves.

Table 4.5 shows the observed and simulated average annual runoff in the calibration and validation periods, together with the percentage error (note that the model fits were adjusted so that observed and simulated annual runoff in the calibration period were not too different). The percentage error in average annual runoff during the calibration period was kept to below 5% for nearly all the catchments. The largest error occurs in the Eden catchment, and is probably due to an overestimation of catchment average rainfall. Errors in estimated average annual runoff are generally higher during the validation period, but are less than 10% for nearly all catchments. Runoff is underestimated in 14 out of the 21 catchments.

Figure 4.9 shows observed and simulated mean monthly runoff (in millimetres) for six of the study catchments over the validation period.

The model simulates mean monthly flow regimes reasonably well. There are a few instances of over- and under-estimation, mostly in spring or autumn. The largest errors are in the Turkey Brook catchment, a responsive, partly urbanised catchment.

TABLE 4.4 Parameters of fitted model, with efficiency criterion calculated over calibration and validation periods

River	C_{max}	S_{max}	K_b	G_{rout}	S_{rout}	b	Efficiency calibration	validation
Don	246.0	140.7	1.701	2.455	0.378	0.749	0.742	0.680
Almond	133.2	92.1	0.621	2.336	0.539	0.447	0.717	0.702
Lyne Water	246.8	195.9	1.011	316.105	0.450	0.260	0.623	0.730
Bedburn Beck*	100.0	40.0	1.000	3.000	0.500	1.500	0.572	0.596
Greta*	91.1	60.0	0.514	5.425	0.617	0.518	0.600	0.543
Dove*	182.3	70.0	1.526	1.500	0.401	1.605	0.753	0.736
Harpers Bk*	150.0	120.0	0.400	3.200	0.600	0.250	0.660	0.581
Wensum	127.7	93.6	0.801	1.053	0.300	0.364	0.834	0.696
Colne	120.0	104.9	0.248	0.556	0.401	0.144	0.562	0.646
Turkey Bk	170.9	111.6	0.472	0.009	0.632	0.531	0.712	0.737
Thames*	350.0	180.0	0.700	10.000	0.200	0.944	0.769	0.809
Lambourn*	420.0	380.0	1.100	2.000	0.100	0.105	0.831	0.758
Medway	211.7	162.7	0.675	3.217	0.542	0.301	0.712	0.747
Avon*	210.0	160.0	2.000	2.000	0.200	0.313	0.817	0.825
Tamar	191.5	150.3	0.569	5.483	0.355	0.274	0.783	0.790
Teme	140.4	121.7	0.304	2.220	0.212	0.153	0.559	0.627
Cynon	270.3	157.3	1.640	6.407	0.557	0.719	0.806	0.824
Ewenny	183.0	107.8	2.411	2.498	0.547	0.697	0.789	0.760
Conwy	103.7	49.2	1.348	14.327	0.639	1.105	0.757	0.727
Eden*	80.0	68.5	0.836	4.638	0.534	0.167	0.458	0.455
Nith	87.9	56.4	1.124	11.631	0.570	0.559	0.715	0.737

Notes: * indicates manually-adjusted parameters. The parameter b is calculated from c_{max} and s_{max}.

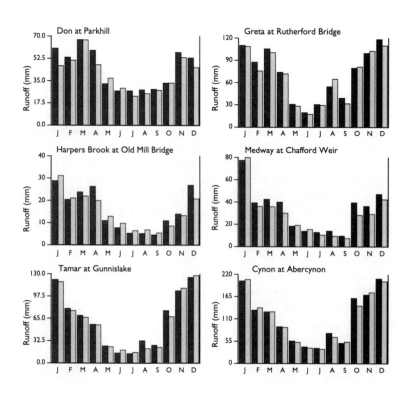

FIGURE 4.9 Observed and simulated monthly mean runoff, validation period — observed: dark, simulated: light

TABLE 4.5 Observed and simulated average annual runoff (mm)

Catchment	Calibration period			Validation period		
	Obs	Sim	% error	Obs	Sim	% error
Don	500	485	-3.0	529	487	-7.9
Almond	553	576	4.2	580	543	-6.4
Lyne Water	570	551	-3.3	572	581	1.6
Bedburn	521	517	-0.8	545	503	-7.7
Greta	907	885	-2.4	834	805	-3.5
Dove	694	664	-4.3	619	594	-4.0
Harpers Bk	199	196	1.5	184	172	-6.5
Wensum	277	276	-0.4	252	244	-3.2
Colne	158	158	0.0	153	153	0.0
Turkey Bk	179	177	-1.1	164	164	0.0
Thames	298	294	-1.3	260	285	9.6
Lambourn	270	268	-0.7	225	224	-0.4
Medway	390	391	0.3	382	343	-10.2
Avon	367	369	0.5	323	337	4.3
Tamar	825	792	-4.0	733	715	-2.5
Teme	457	420	-8.1	398	395	-0.8
Cynon	1512	1562	3.3	1362	1335	-2.0
Ewenny	1086	1010	-7.0	1046	947	-9.5
Conwy	1865	1957	4.9	1753	1640	-6.4
Eden	770	858	11.4	720	796	10.6
Nith	1245	1307	5.0	1185	1277	7.8

Obs = observed, Sim = simulated

Similar patterns are found in the both calibration period and the validation period. There is a slight tendency for greater errors in the validation period, but differences are small.

Figure 4.10 shows the observed and simulated daily flows for two example years for two of the study catchments. 1982 is during the calibration period and 1987 falls within the validation period. The graphs also show bias, calculated as simulated minus observed flow. Figure 4.11 shows observed and simulated flow duration curves for the same catchments, again for the calibration and validation periods.

The model simulates daily flow regimes less well than monthly flow regimes. It reproduces the general regime pattern reasonably well, but details are occasionally poorly simulated: the model does however produce regimes that vary between catchments. Several general conclusions can be drawn from the daily flow regime diagrams.

The model nearly always underestimates high flow peaks (this is a failing common to nearly all daily rainfall-runoff models). Some peaks are missed entirely, but in most cases peaks exist but are too small. The model is particularly bad at simulating high flows in summer and autumn. There are two possible reasons why high flows are poorly simulated. Firstly, the model may not simulate the runoff-generation process very well: too little of the catchment is assumed to be saturated and

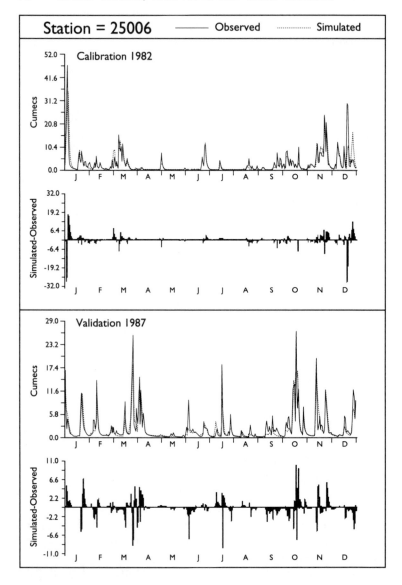

FIGURE 4.10 Observed and simulated daily flows for two years at two case study catchments

able to respond rapidly to rainfall. Secondly, the model uses catchment average daily rainfall as input data. In practice, rainfall may be highly localised in one part of the catchment (although most of the study catchments are small) or might have fallen in just a few hours. A shorter time step might have led to better simulations of high flows, but data were not available.

In some catchments in the north and west both the timing and magnitude of flood peaks appears to be wrong. This is not due to errors in the model but to errors in the treatment of snowfall and snowmelt. Visually, there is little obvious difference in performance between the example calibration

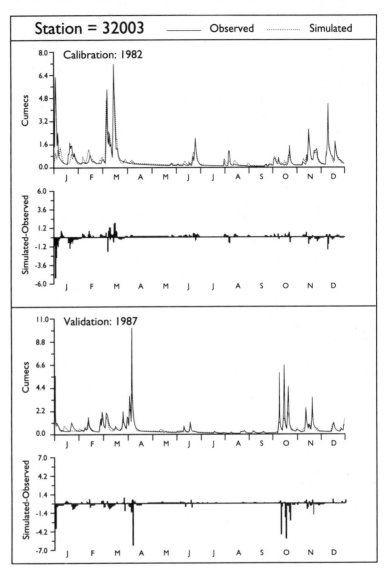

FIGURE 4.10 (cont.) Observed and simulated daily flows for two years at two case study catchments

and validation years. In some cases the fit is worse in the validation year (e.g. the Don); in other cases it is rather better. The daily flow regimes in groundwater dominated catchments (the Lambourn and the Avon) are poorly simulated. The high frequency day-to-day fluctuations are not reproduced at all: the model misses what little rapid runoff occurs — usually from the limited saturated areas next to the river course.

Assessment

The evaluation of the rainfall-runoff model showed that mean monthly flow regimes are well simulated, that the general pattern of daily flow regimes is well reproduced (with the exception of high flow peaks), and that the flow duration curves from simulated and observed data are quite similar, except at extreme high and low flows. There is a general tendency to underestimate low flows, particularly when flow is less than the observed flow exceeded 95% of the time. The model generally fits the groundwater-dominated catchments less well than it fits more responsive catchments.

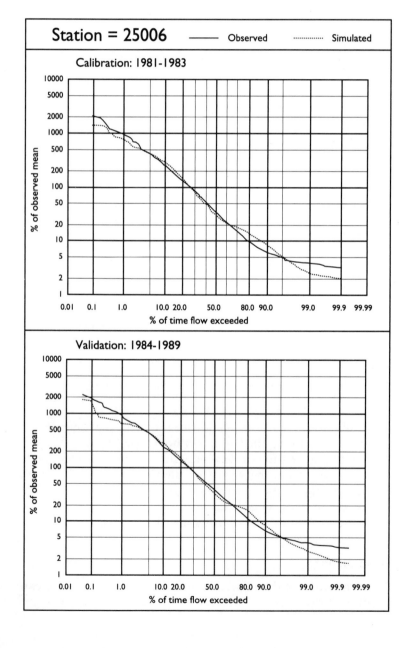

FIGURE 4.11 Observed and simulated flow duration curves for two catchments

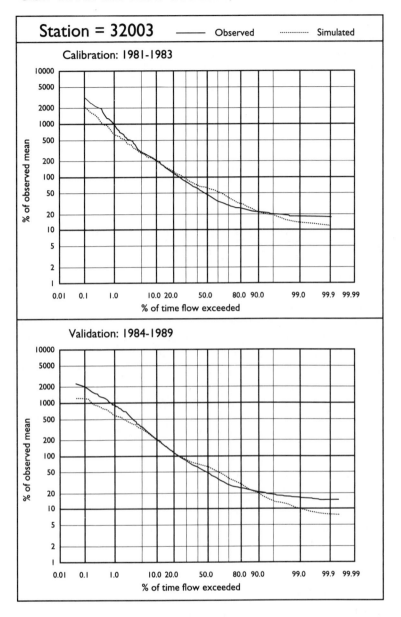

FIGURE 4.11 (cont.) Observed and simulated flow duration curves for two catchments

Applying the model to gridded input data

The daily rainfall-runoff model was not only applied to individual catchments, but was also applied across the whole of Britain at a spatial resolution of 0.5 × 0.5° (approximately 2500 km^2). It was assumed that each 0.5 × 0.5° cell represented a catchment, and the same model parameters were assumed for each cell. These parameters are shown in Table 4.6, and were selected both to be representative of the range of parameters found in the study catchments and to produce estimates of average annual runoff across Britain which matched the observed spatial variability. The values of the two routing parameters G_{rout} and S_{rout} are not important, because the analysis of the gridded model output focused on average annual runoff. No adjustments for snowfall or snowmelt were made when the model was applied to gridded data.

Parameter	C_{max}	S_{max}	K_b	G_{rout}	S_{rout}
Value	200	170	1	1	0.5

TABLE **4.6** Parameters of rainfall-runoff model applied to gridded climate data

CLIMATE CHANGE SCENARIOS FOR THE GREAT BRITAIN CASE STUDY

This section reviews the baseline data and climate change scenarios developed for the Great Britain case study. The study defined changes in monthly temperature, precipitation and potential evapotranspiration from GCM output, and applied these changes to observed historical baseline daily time series. The study used observed historical daily data rather than stochastically-generated data, because it was not possible to create a stochastic daily precipitation model that reproduced sufficiently realistically the temporal characteristics of observed time series; dry spells tended to be too short.

The baseline climate data

The period 1961 to 1990 was used as the baseline period, and time series of daily rainfall, potential evapotranspiration and temperature (where necessary for the snow model) were created for each catchment.

Daily temperature data were available for 1980 to 1990 (and were used in the model development, calibration and validation), but were not available for the period back to 1961. Catchment daily temperature data for the baseline period were therefore estimated from the daily central England temperature record (Manley, 1974 and updates) using catchment regression relationships derived for each month from the 1980 to 1990 data. The relationships were generally good, with coefficients of determination in excess of 80%, except for the Don in north east Scotland: here the relationship between local temperature anomalies and central England anomalies is weakest.

Monthly potential evaporation data for each catchment for the period 1961 to 1990 were taken from the MORECS data base (Thompson *et al.*, 1981). MORECS potential evaporation is calculated using the Penman-Monteith formula, assuming a grass cover. No corrections were made to adjust the potential evaporation for a given catchment on the basis of land cover. Daily potential evaporation was derived simply by dividing the monthly total equally amongst all the days in a month.

Daily catchment average rainfall data for each catchment for 1961 to 1990 were extracted from archives at the Institute of Hydrology. The main complication was that the numbers of rain gauges contributing to the catchment average varied over time, introducing potential inhomogeneities into the record. For each catchment, a time series of catchment average *monthly* rainfall was therefore created over the period 1961 to 1990, extending records where necessary by correlation with nearby long-duration rainfall records. The catchment daily rainfall data were then rescaled, month-by-month, to create a homogeneous time series of daily rainfalls. The approach assumes that whilst the volumes of the catchment average daily rainfall in the original daily series may be unreliable, the daily series gives a reliable indication of the *proportional* distribution of catchment average daily rainfall through the month. This is unrealistic where one daily rain gauge is used to estimate catchment average rainfall over a large catchment; the final daily series will contain too many dry days.

When the hydrological model was applied at the 0.5 × 0.5° resolution across Britain, baseline data were derived from a gridded data set produced by the Climatic Research Unit (Hulme et al., 1995). This data set includes monthly precipitation and temperature over the period 1961 to 1990, together with monthly mean vapour pressure, sunshine hours and windspeed. Daily rainfall for each grid cell over the period 1961 to 1990 was determined from the monthly rainfall in the following way. First, all the rain gauges falling within a 0.5 × 0.5° cell were identified, and each day's rain expressed as a percentage of the monthly total for that rain gauge. A cell average time series was then created by simply averaging on each day all the scaled daily rainfalls, and the resulting cell time series was rescaled to ensure that each monthly total equalled 100%. For each cell, a time series of daily rainfall from 1961 to 1990, in millimetres, could then be created by combining the cell monthly rainfall totals with the cell daily percentage time series. Cell monthly potential evaporation was estimated from the monthly average temperature, net radiation (determined from sunshine hours), windspeed and humidity using the Penman-Monteith formula: the 1961 — 1990 average values were used each year.

Introduction to the climate change scenarios

The climate change scenarios used in the study are based on those created for the 1996 Climate Change Impacts Review Group (CCIRG, 1996; Hulme, 1996). The scenarios were constructed by the Climatic Research Unit at the University of East Anglia from output from the Hadley Centre transient climate change experiment (UKTR: Murphy, 1995; Murphy and Mitchell, 1995), essentially in four stages.

The first stage involved the selection of an emissions scenario. In 1992 the IPCC defined a set of emissions scenarios, representing different assumptions about trends in emissions over time (Figure 3.3). Some of these scenarios were subsequently modified for the 1996 IPCC Scientific Assessment (IPCC, 1996), to allow for the more rapid phase-out of CFCs and HCFCs than was initially assumed: these scenarios are known as the "augmented IS92 scenarios". The CCIRG climate change scenarios are based on the augmented IS92a emissions scenario, which falls in the middle of the six IS92 scenarios.

The second stage was the conversion of this emissions scenario into a global warming scenario, using a simple one-dimensional upwelling-diffusion energy balance model (MAGICC: Hulme *et al.* 1995b). Figure 4.12 shows the global temperature change under the augmented IS92a scenario, assuming a climate sensitivity of 2.5°C (this is the assumed equilibrium response of the atmosphere to a doubling of equivalent CO_2 concentrations, and is the IPCC's best guess). During the 2020s, the global average temperature would be 0.92°C higher than the 1961 to 1990 baseline; by the 2050s, the temperature increase would be 1.63°C. These changes do not include the possible effects of sulphate aerosols (Chapter 2).

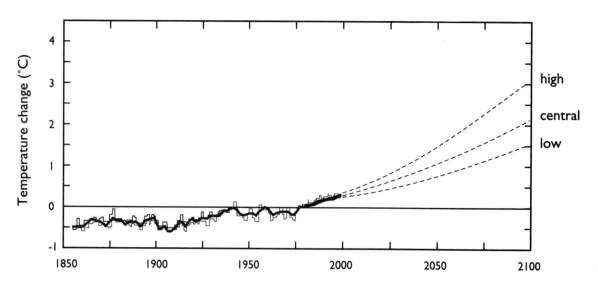

FIGURE 4.12 Global temperature changes under emissions scenario IS92a (augmented) and a climate sensitivity of 2.5°C. Figure provided by Dr M Hulme, Climatic Research Unit, University of East Anglia.

The third stage defined the geographic pattern of change using the results from the Hadley Centre transient experiment (UKTR). This experiment assumed a 1% compound increase in equivalent CO_2 concentrations from 1990 levels over a 75-year period. The GCM has a resolution of 2.5 × 3.75°, and produces a reasonable simulation of current climate patterns (Hulme, 1994b). The geographic pattern of change was taken from the last simulated decade (model years 66 to 75), by which time the simulated global temperature had increased by 1.76°C, relative to the baseline. The patterns of change for each variable and month were scaled according to the warming modelled under the augmented IS92a scenario (for example, by 1.63/1.76 for the 2050s). This approach assumes that the spatial pattern of change remains constant over time, which will not in fact be true. Also, a different set of model years would give a different pattern of change.

The resolution of the UKTR simulations, although good relative to other global climate models, is still coarse, and many of the small-scale influences on the British climate, such as local topography and the effects of the surrounding seas, cannot be simulated. Spatial patterns of change therefore reflect changes in large-scale climate systems, and tend to have a north-south dimension. Possible changes in the west-east gradient across Britain cannot be determined at this spatial scale. The spatial resolution is also too coarse for the definition of catchment-scale scenarios. Chapter 3 outlines several possible methods for "downscaling" to the catchment scale. The simplest approach was used in the study, and the coarse resolution changes were simply interpolated by the Climatic Research Unit onto a 0.5° × 0.5° grid as the fourth stage in the scenario definition process. This high resolution detail is largely spurious and wholly statistical: it is controlled by the large-scale changes.

Changes in temperature

Figure 4.13 shows the changes in mean temperature by the 2050s for winter (December to February), spring (March to May), summer (June to August) and autumn (September to November). By the 2050s, winter temperature rises by 2°C in the south-east, and by around 0.8°C in the north. In summer, the range is between 1.8 and 1.2°C. The increase in the north is lower than in the south — contrary to that assumed in other scenarios (e.g. CCIRG, 1991) — because the transient simulation experiment captures the slow response of the north Atlantic to increasing temperatures.

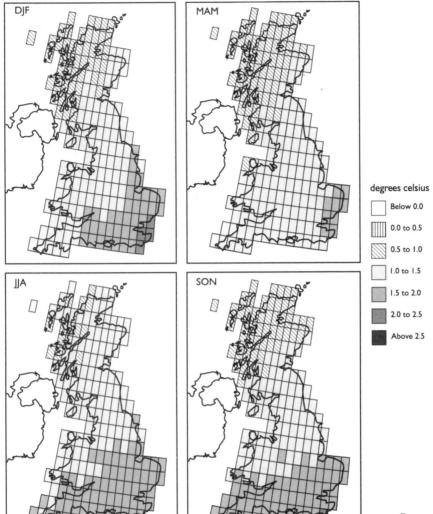

degrees celsius

Below 0.0

0.0 to 0.5

0.5 to 1.0

1.0 to 1.5

1.5 to 2.0

2.0 to 2.5

Above 2.5

FIGURE 4.13   Change in seasonal temperature by the 2050s

Changes in precipitation totals

At the annual scale, precipitation is increased by the 2050s across the whole of Britain, with increases ranging from less than 5% in the south to over 15% in the north. There is greater regional variation in seasonal and monthly rainfall changes. Figure 4.14 shows percentage change in seasonal precipitation totals across Britain, and Figure 4.15 shows monthly changes for eight locations (Figure 4.2). Precipitation increases throughout Britain in winter, with the greatest percentage increase in the south (up to 10%). In summer, precipitation increases in northern Britain by up to 10%, but decreases south of around 53°N (near Nottingham). In the south-east, the reduction in summer rainfall totals is approximately 10%.

The spatial pattern of precipitation changes is particularly sensitive to the period used to define the change scenarios from the transient model output.

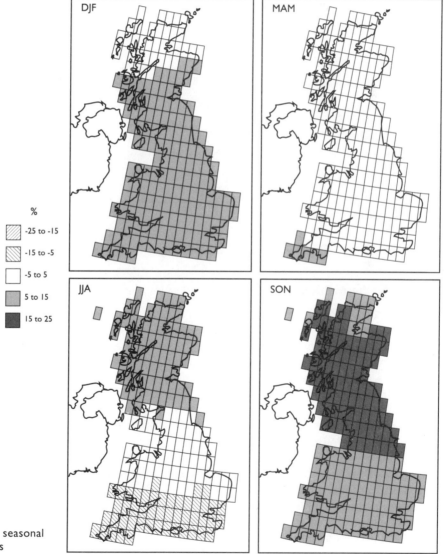

FIGURE 4.14   Change in seasonal precipitation by the 2050s

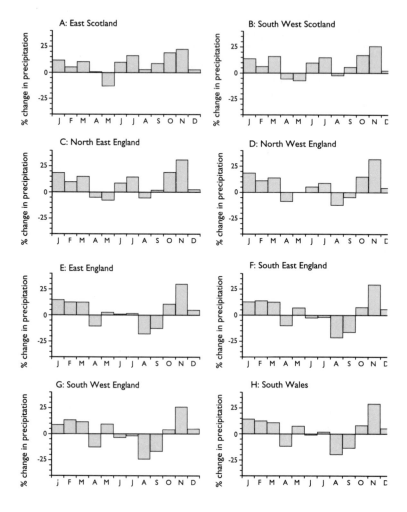

FIGURE 4.15   Change in monthly precipitation by the 2050s at eight locations

Changes in the number of rain-days

It is difficult to define changes in daily rainfall from a global climate model (Chapter 3) because the spatial scale is too coarse: each day's simulated weather represents an average over a very large area. In the construction of the 1996 CCIRG scenarios, a rain-day was therefore defined as occurring when at least 2mm of rainfall fell in a ($2.5 \times 3.75°$) grid cell. This threshold value was selected because it gives similar mean rain-day frequencies over Britain from climate model output to observed climatologies (Hulme *et al.*, 1994).

The number of days on which rain falls increases in winter throughout Britain, although generally by less than one day per month. In summer, rain falls more frequently in the north and less often in the south. although again changes are small. The proportional changes in rain-days are generally less than changes in precipitation totals, implying an increase in average precipitation intensity.

Changes in potential evapotranspiration

An increase in temperature means that air is able to hold more moisture, and by itself would cause evaporation to increase. Evaporation rates are also determined by the amount of energy available: although in general this is likely to increase, net radiation might be reduced in some seasons if cloud cover increases. Evaporation is also constrained by atmospheric humidity: although higher temperatures will tend to decrease humidity, atmospheric moisture content will also increase — because of increased evaporation — and locally an increase in atmospheric moisture may outweigh the effects of an increase in temperature. This is particularly likely to be the case where prevailing winds pass across large areas of warmer ocean. Changes in windspeed will also affect evaporation rates, and so will changes in plant physiological properties (specifically the number and characteristics of stomata) and the vegetation in a catchment (Chapter 3).

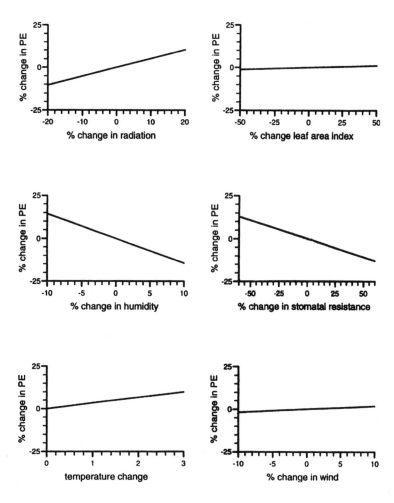

FIGURE 4.16 Sensitivity of Penman-Monteith potential evaporation in southern Britain to changes in meteorological and plant characteristics (Reynard, 1993)

A sensitivity analysis using data from Heathrow Airport with the Penman-Monteith formula (Arnell and Reynard, 1993; Reynard, 1993) showed that evaporation rates in Britain were highly sensitive to changes in humidity in particular (Figure 4.16). This contrasts with the results of Martin *et al.* (1989) from the mid-west of the USA (Chapter 3), which showed the greatest sensitivity to changes in radiation. In humid Britain the level of atmospheric humidity is a significant constraint on evaporation, whereas in the mid-west humidity is rarely a constraint, so changes have relatively little effect.

The 1996 CCIRG scenario was based on the Penman evaporation equation, and assumed no change in plant

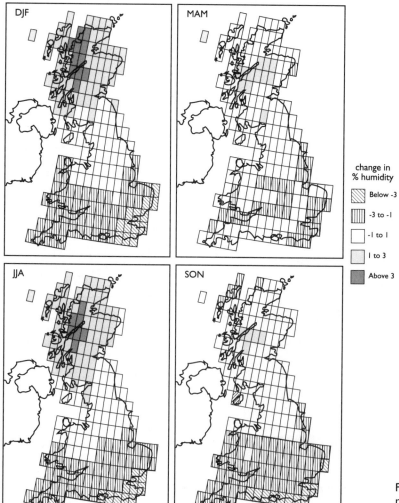

FIGURE 4.17 Percentage change in relative humidity by the 2050s, by season

physiological properties and, at the catchment scale, no change in vegetation composition. The scenario incorporated changes in temperature, humidity, net radiation and windspeed: changes in temperature and humidity were the most important.

Absolute humidity increases everywhere across Britain, throughout the year. *Relative* humidity, however, is affected by air temperature, and the rise in air temperature means that across most of Britain relative humidity declines (Figure 4.17). In northern Scotland, however, relative humidity increases, because the temperature effect is not enough to offset the increase in absolute humidity. Net radiation generally increases in summer — because skies are clearer — and shows little change in winter (Figure 4.18).

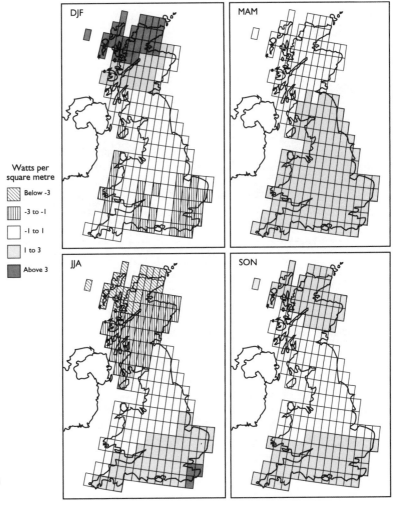

FIGURE 4.18 Change in net radiation (Wm²) by the 2050s, by season

Windspeed showed a slight increase throughout the year. In winter, the greatest increase, up to 7%, is in the south, whilst during summer increases generally are smaller, around 2 to 3%.

Figure 4.19 shows seasonal changes in potential evaporation, resulting from the assumed changes in inputs, and Figure 4.20 shows the monthly percentage changes for eight locations. The largest percentage increases are in winter, but the absolute increases here are small because the initial amounts are low. Potential evaporation *declines* across much of northern Scotland, because humidity increases and, in some seasons, net radiation is reduced. Figure 4.21 shows the annual change in potential evaporation across Britain, both as a percentage and in millimetres.

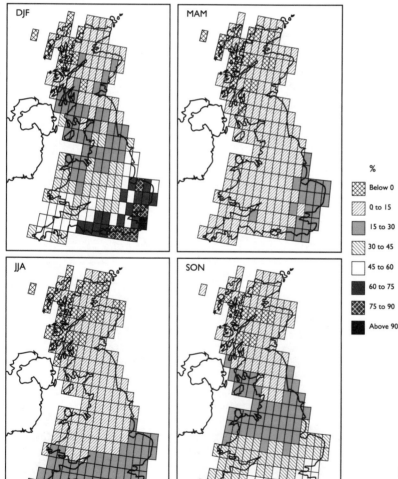

FIGURE 4.19 Percentage change in seasonal potential evaporation by 2050

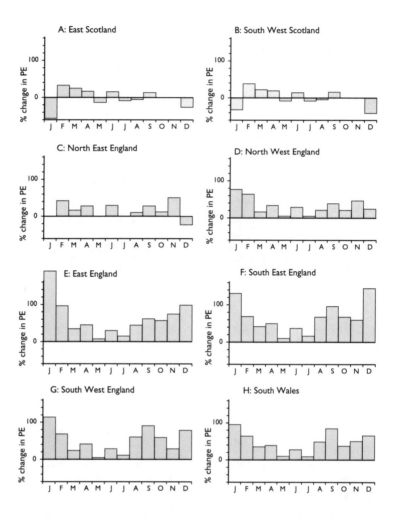

FIGURE 4.20 Change in monthly potential evaporation by 2050 at eight locations

The CCIRG potential evaporation scenario shows a greater percentage change in potential evaporation than that used in Belgium and Switzerland by Bultot *et al.* (1988a; 1992). It is similar, in percentage terms, to the EVP2 scenario of Arnell *et al.* (1990) and Arnell (1992), and close to the PE2 scenario as used by Arnell and Reynard (1993). This latter scenario was based on the application of the Penman-Monteith equation with assumed changes in temperature, humidity, net radiation and windspeed.

Application of scenarios

The scenarios were applied to the 30-year baseline daily time series to create perturbed data. The temperature changes were applied as absolute increases, whilst all other changes were expressed in percentage terms. A given monthly change was applied to each day in the month.

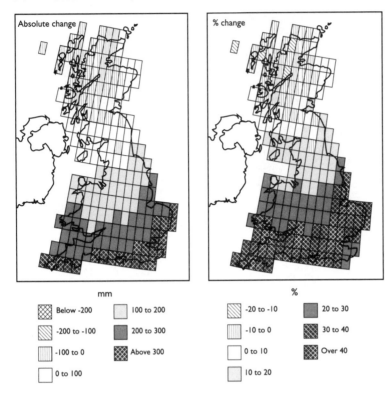

Absolute change

% change

mm

%

⊠ Below -200	□ 100 to 200	⊠ -20 to -10	■ 20 to 30
⊠ -200 to -100	■ 200 to 300	▦ -10 to 0	⊠ 30 to 40
▦ -100 to 0	⊠ Above 300	□ 0 to 10	⊠ Over 40
□ 0 to 100		□ 10 to 20	

FIGURE 4.21 Absolute and percentage changes in annual potential evaporation by 2050

Several studies (e.g. Bates *et al.*, 1994) have generated daily time series stochastically, using a model of daily weather characteristics. This approach was not used here, because it was not possible to create a simple generator which reproduced well observed daily rainfall characteristics. Two-state first-order Markov models, as used in most other studies, tended to generate dry spells that were too short, and simulated river flows would therefore be inaccurate.

The use of observed, rather than synthetic, data, however, made it rather difficult to incorporate the effects of changes in the number of days on which rain falls. An objective procedure was developed for increasing or decreasing the number of days by *n* consistently. Increasing the number of rain-days involved the following stages:

1. Apply the monthly percentage change to every day in the month.
2. Calculate the mean rainfall on rain-days.
3. Select randomly *n* of the dry days.
4. Assign these new rain-days a rainfall equal to the mean daily rainfall.
5. Reduce the rainfall on the original rain-days, in proportion to the daily rainfall.

Reducing the number of rain days was slightly simpler:

1. Apply the monthly percentage change to every day in the month.
2. Select randomly *n* of the rain-days, and set their rainfall to zero.
3. Distribute the rainfall which originally fell on these days over the remaining rain-days, in proportion to the rainfall on the rain-day.

These procedures are arbitrary, but can be objectively applied. The main problem is that there is no control over which days are selected to become wet or dry. A refinement might be to select only from amongst the days following a wet day.

The core CCIRG scenario represents average conditions with a stable climate similar to that of the 2050s.

Changes in variability and extremes

The scenarios make no explicit reference to changes in year-to-year variability, and are restricted to assumptions about changes in the mean. Whe the scenarios are applied as a percentage rescaling to each value as for precipitation or potential evaporation — the effect is to alter both the mean and the standard deviation, keeping the relative variability, as expressed by the coefficient of variation, and higher moments such as skewness the same. When the change is an addition — as for temperature — the effect is to alter the mean but keep the standard deviation the same, although skewness and higher moments are unaffected. In reality global warming might lead to differing changes in the mean, variability and higher moments, and the implied assumptions of constant coefficient of variation (for precipitation and potential evaporation) or standard deviation (for temperature) may not be appropriate. However, no scenarios for changes in inter-annual variability have yet been produced from climate model simulations, because the sampling variability around estimates of variability from different periods is large.

Some indications of changes in the frequency of occurrence of extremes can be obtained, however, by making simple changes to statistical distributions. A shift in the mean or standard deviation of a statistical deviation means that defined values are exceeded with a different probability, and small changes in the characteristics of a distribution can give large changes in the frequency of exceedance of fixed values. For example, under the CCIRG scenario, the return period of the 1995 summer temperature would decrease from once in 80 years to once in three years (CCIRG, 1996): this assumes just a change in the mean.

Transient change scenarios

Most of the assessments in the Great Britain case study assumed a stable "equilibrium" climate with an average similar to that expected in the 2050s. Some simulations (presented in Chapter 6) used transient scenarios, assuming a sequence of changes over the period 1990 to 2050. The transient scenarios were created by rescaling the standardised UKTR changes by the global average temperature change for each year between 1990 and 2050, where the annual global change was determined by interpolating linearly between 1990 (0°C change) and 2050 (1.63°C change). This assumes not only that the pattern of change does not vary with time, but also that the change in regional climate is linear. Neither of these assumptions is likely to be true, but the approach does enable a first estimate of transient changes in hydrological regimes. The rate of temperature change varies across Britain from 0.2 to 0.33°C per decade — close to the global value — and annual precipitation changes by up to 1.7% per decade in both winter and summer.

The transient scenarios were applied by doubling the 30-year baseline to create a 60-year time series, and then perturbing each year to simulate change over the period 1990 to 2050. This approach retains and repeats the temporal structure of the year-to-year variability in the original 30-year series, but was necessary because the stochastic generation of daily climate data had been eliminated. An alternative approach — randomly sampling from the individual years to create a 60-year series — was not pursued.

Uncertainties and comparisons with other scenarios

The CCIRG scenario clearly has several degrees of uncertainty. In increasing order of uncertainty, these are

- *Uncertainty in the emissions scenario.* The IPCC produced several scenarios, making different assumptions about rates of economic and population growth. However, for the first few decades, there is little difference between these scenarios.

- *Global temperature change.* The main uncertainty in estimating the effects of a given emissions scenario on temperature lies in selection of the "climate sensitivity", or the assumed stable equilibrium effect of a doubling of CO_2 concentrations. The CCIRG scenario uses the IPCC "best guess": the range of uncertainty by the 2050s is between 1.1°C and 2.4°C (CCIRG, 1996).

- *Regional temperature change.* Patterns of regional temperature change are dependent on the climate model used. As indicated below, different models give different patterns of change.

- *Regional precipitation change*. Regional precipitation changes too are dependent on the climate model used, and differences between models are greater than for temperature.

- *Regional change in potential evaporation*. Changes in regional potential evaporation are a function of changes in several climatic variables.

The CCIRG scenario is therefore not to be seen as a forecast, but as one feasible future.

Different climate models produce slightly different patterns of change, and these differences are much greater for precipitation than for temperature. In terms of temperature, the biggest differences are between the patterns simulated by equilibrium experiments (comparing $2 \times CO_2$ with $1 \times CO_2$ conditions) and those simulated in transient experiments which use a coupled ocean-atmosphere model. In the earlier, un-coupled experiments, temperature increases were at their maximum at high latitudes, and northern Britain warmed by a greater amount than southern Britain. In the coupled transient experiments, however, the oceans take a long time to respond to increased temperatures, so parts of the world with climates strongly influenced by the sea show a smaller increase in temperature than continental regions. In transient experiments, the North Atlantic tends to warm very slowly, so temperatures in northern Britain rise by a smaller amount than temperatures in southern Britain.

Virtually all climate models simulate an increase in winter rainfall across most of Europe, although the amount of increase varies. Most models also simulate an increase in summer rainfall in northern Europe and a decrease in southern Europe. The dividing line between increase and decrease varies from the centre of France in some models to the north of Scotland in others. In the UKTR experiment, the dividing line passes through central England. Compared with other climate models, the UKTR experiment generates a relatively wet response to global warming over Britain, and a relatively subdued temperature rise (CCIRG, 1996).

A set of climate change scenarios was developed in 1991 from a range of climate models (CCIRG, 1991). These earlier scenarios assumed an increase in winter rainfall — with a best guess of 8% by 2050 — and no change in summer rainfall; they gave no indications of spatial variability in change. A temperature increase of 2.1°C in summer was assumed, with increases in winter temperature of 2.3°C in southern Britain and 2.9°C in the north. The 1996 CCIRG scenario therefore assumes a smaller increase in temperature and, significantly, a different spatial pattern. The 1991 CCIRG scenario made no reference to changes in potential evaporation, but Arnell and Reynard (1993) defined a set. One (PE1) assumed just a change in temperature, and resulted in an increase in annual potential evaporation of around 9% by the 2050s. PE2 made further assumptions about changes in humidity, windspeed and radiation, and assumed an annual increase of around 30%. PE3 made further assumptions about changes in plant properties, and gave an annual increase of approximately 20%. The 1996 CCIRG scenario is closest to PE2 (Arnell and Reynard, 1993).

A final uncertainty arises from the omission of the effects of sulphate aerosols in the CCIRG scenario. As summarised in Chapter 2, sulphate aerosols counteract the enhanced greenhouse effect by reflecting incoming solar radiation back into space. They tend therefore to reduce the simulated increase in temperature, and at the global scale sulphate aerosols could suppress global warming by 0.3°C by 2050 (under the IS92a scenario: Wigley and Raper, 1992). At the regional scale, however, the effect of sulphate aerosols depends on their spatial distribution. There have been few simulation studies using transient climate models (Mitchell *et al.*, 1995a;b.): the effects on rainfall and potential evaporation are unclear, and so are future actions to curb emissions of sulphate aerosols.

The following chapters explore the consequences of the 1996 CCIRG scenario for hydrological regimes in Britain. It must be remembered throughout that different scenarios would give different impacts, and results from other studies will frequently be introduced and compared.

CHAPTER 5

Changes in water resources in Britain

This chapter explores the changes in river flows and groundwater recharge in Britain that might occur by the 2050s. It is based around the catchments and scenarios introduced in Chapter 4. The simulated changes are placed in the context of changes identified by other studies both in Britain and other environments.

THE ANNUAL WATER BALANCE

Changes in the annual water balance

Many sensitivity studies have examined the effects of defined arbitrary changes in precipitation, temperature and potential evapotranspiration on runoff. Some of these studies (e.g. Revelle and Waggoner, 1983; Arnell and Reynard, 1989) used empirical statistical models estimating average annual runoff from annual precipitation, temperature and evaporation, but the estimated sensitivities to change are very dependent on the form of statistical model used. Other studies have used water balance models, operating at monthly, daily or finer time steps, and the results of such studies are emphasised here.

There are two general conclusions from the published sensitivity studies:

- annual runoff volume is more sensitive to changes in precipitation than to changes in potential evapotranspiration;
- a given percentage change in precipitation results in a greater percentage change in runoff.

With this amplification (Schaake, 1990) increasing as the proportion of precipitation going to runoff decreases: arid catchments show greater sensitivity to change than humid catchments. This amplification arises simply because the percentage change in precipitation minus evaporation (P-E) will be greater than the percentage change in P, with the amplification increasing as E approaches P.

Wigley and Jones (1985) showed that the sensitivity of annual runoff to changes in precipitation and actual evaporation could be estimated from

$$\frac{R_1}{R_0} = \frac{\alpha - (1-\gamma_0)\beta}{\gamma_0}$$

where R is annual runoff, α is the fractional change in annual rainfall, β is the fractional change in actual evaporation, γ is the ratio of annual runoff to annual rainfall (the runoff coefficient), and the subscripts 0 and 1 represent current and changed conditions respectively. The fractional change in runoff, R_1/R_0, is clearly dependent on the runoff coefficient, and increases as γ_0 decreases. However, this expression is only valid if the fractional change in annual actual evaporation β is independent of the change in annual rainfall α, and this is not necessarily the case.

Table 5.1 summarises the sensitivity of annual runoff to fixed changes in rainfall and a temperature increase of 2°C, applied equally throughout the year, as simulated for several catchments using water balance models (note that the change in potential evaporation associated with a 2°C rise in temperature varied between the catchments tabulated, both because the initial climatic conditions were different and because different methods were used to estimate evaporation). The greatest changes are in the catchments with the lowest runoff coefficients, namely the Pease in Texas, Nzoia in Kenya and Saskatchewan in Canada. Figure 5.1 shows "sensitivity surfaces" for three of the catchments, plotting change in runoff against change in temperature and precipitation (note that the interpolated surfaces are based on few points). In all catchments, an increase in annual precipitation of 10% is enough to offset the higher evaporation associated with a 2°C rise in temperature. In humid catchments the relationships between precipitation change and runoff change tend to be linear, but in drier catchments the amplification ratio increases as the precipitation increase becomes larger. The slope and the spacing of the lines define the sensitivity of annual runoff to change in rainfall and evapotranspiration. The steeper the lines, the greater the sensitivity to change in rainfall relative to change in evapotranspiration; the closer the spacing between the lines, the greater the relative effect of a change in climate on annual runoff.

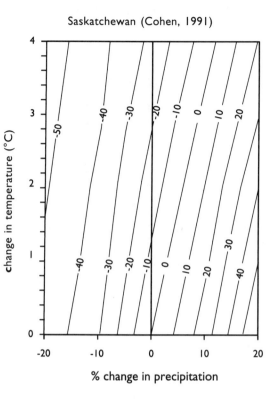

Saskatchewan (Cohen, 1991)

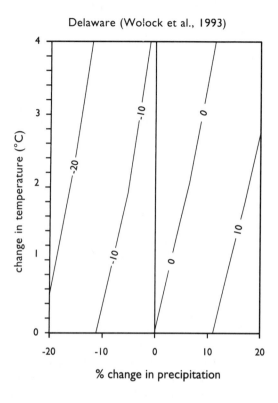

Delaware (Wolock et al., 1993)

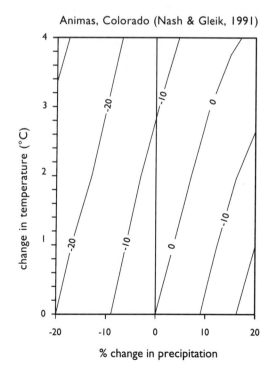

Animas, Colorado (Nash & Gleik, 1991)

FIGURE 5.1 Sensitivity of annual runoff to changes in precipitation and temperature: Cohen, 1991; Wolock *et al.*, 1993 and Nash and Gleick, 1991

TABLE 5.1 Percentage change in annual runoff with different changes in annual precipitation and an increase in temperature of 2°C

River	Location	Reference	Percent change in precipitation				
			-20	-10	0	10	20
White	Colorado basin, USA	Nash & Gleick (1991)	-23	-14	-4	7	19
East	Colorado basin, USA	Nash & Gleick (1991)	-28	-19	-9	1	12
Animas	Colorado basin, USA	Nash & Gleick (1991)	-26	-17	-7	3	14
Upper Colorado	Colorado basin, USA	Nash & Gleick (1991)		-23	-12	1	
Lower Delaware	USA	Wolock et al (1993)	-23		-5		12
Saskatchewan	Canada	Cohen (1991)	-51	-28	-15	11	40
Pease	Texas, USA	Nemec & Schaake (1982)	-76	-47	-12	40	100
Leaf	Mississippi, USA	Nemec & Schaake (1982)	-50	-30	-8	16	40
Nzoia	Kenya	Nemec & Schaake (1982)	-65	-44	-13	17	70
Mesohora	Greece	Mimikou & Kouvopoulos (1991)	-32	-18	-2	11	25
Pyli	Greece	Mimikou & Kouvopoulos (1991)	-25	-13	-1	13	25
Jardine	North East Australia	Chiew et al (1995)	-32	-17	-1	11	28
Corang	South East Australia	Chiew et al (1995)	-38	-21	-3	16	33

The change in annual runoff depends on how the change in rainfall and evapotranspiration is distributed through the year. Arnell *et al.* (1990) showed using a monthly water balance model applied to a set of British catchments that the greater the proportion of any increase in rainfall that fell in winter — the primary runoff-generation season — the greater the increase in annual runoff. In one catchment (the Medway), an increase in rainfall of 10% applied in each month would increase annual runoff by 17% (assuming no change in potential evaporation): if the 10% extra rainfall was concentrated in winter, annual runoff would increase by 22%.

TABLE 5.2 Change in annual water balance components in five Belgian catchments and one Swiss catchment (Bultot *et al.*, 1988a; 1992; Gellens, 1993)

Catchment	Percent change in annual total				
	Precipitation	Potential evapo-transpiration	Actual evapo-transpiration	Snowfall	Runoff
Zwalm (Belgium)	7.3	8.6	6.6	-18	9.8
Dyle (Belgium)	6.7	8.7	6.7	-38	6.9
Semois (Belgium)	4.7	9.4	7.8	-46	3.1
Aa (Belgium)	6.7	8.7	10.1	n.a.	10.0
Berwinne (Belgium)	6.0	7.9	6.5	n.a.	8.0
Murg (Switzerland)	4.5	10.0	9.1	n.a.	0.3

Changes in all the components of the annual water balance were calculated by Bultot *et al.* (1988a; 1992) for study catchments in Belgium and Switzerland, using a daily water balance model, and are tabulated in Table 5.2. The details of the changes in annual runoff and evapotranspiration depend on the climate change scenario used and its annual cycle, but it is important to note that percentage changes in actual evapotranspiration are less than percentage changes in potential evapotranspiration: this point is returned to in the Great Britain case study.

THE GREAT BRITAIN CASE STUDY

This section explores the effects of climate change on components of the annual water balance — potential and actual evaporation, snowfall and runoff — in British catchments. Most of the analysis is based on the 1996 CCIRG scenario, but to enable comparisons with other published studies it is interesting first to investigate the sensitivity of the British water balance to arbitrary changes in inputs.

Sensitivity studies

Table 5.3 shows the change in average annual runoff for four of the study catchments, with a 10% increase in potential evaporation and changes in precipitation ranging from a reduction of 20% to an increase of 20%. The changes in precipitation and evaporation were applied equally through the year. The increase in potential evaporation of 10% is slightly higher than the increase which would result from a rise in temperature of 2°C (assuming no change in the other controls on evaporation rates), but the results in Table 5.3 are broadly comparable with those in Table 5.1.

The table illustrates the amplification of changes in rainfall and shows that, in common with other areas, the lower the runoff coefficient, the greater the sensitivity to change. Sensitivities in British catchments are greater than those in the headwaters of the Colorado (Table 5.1), and similar in magnitude to those in Greece.

Figure 5.2 shows the joint effect of changes in precipitation and potential evaporation on annual runoff, for the four catchments in Table 5.3. As in Figure 5.1, the spacing of the lines reflects the sensitivity to change in rainfall, and the slope indicates the relative sensitivity to changes in rainfall and potential evaporation. Clearly, annual runoff in Britain is most sensitive to changes in rainfall.

Catchment	Runoff coefficient	Percent change in precipitation				
		-20	-10	0	10	20
Don at Parkhill	0.56	-32	-19	-5	10	25
Greta at Rutherford Bridge	0.61	-32	-18	-4	10	25
Harpers Brook at Old Mill Bridge	0.31	-44	-27	-9	12	35
Medway at Chafford Weir	0.48	-38	-23	-6	11	29

TABLE 5.3 Percentage change in annual runoff with different changes in annual precipitation and an increase in potential evaporation of 10%.

Table 5.3 and Figure 5.2 provide some indications of the sensitivity of British catchments to changes in climatic inputs, but are not particularly helpful in assessing the possible effects of global warming. Studies based on scenarios are of much greater value.

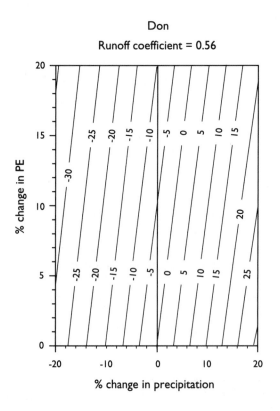

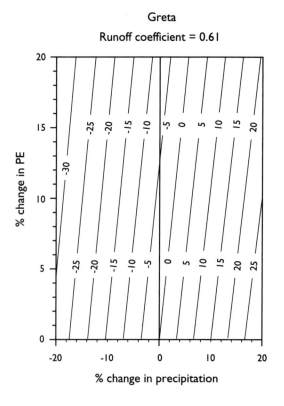

FIGURE 5.2 Sensitivity of average annual runoff in four British catchments to changes in precipitation and potential evaporation: the changes are assumed to apply consistently through the year (continued overleaf).

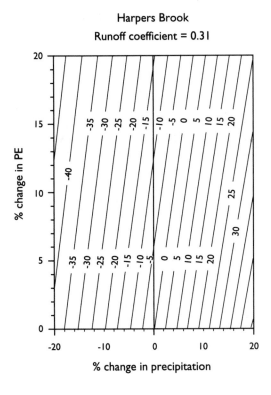

Harpers Brook
Runoff coefficient = 0.31

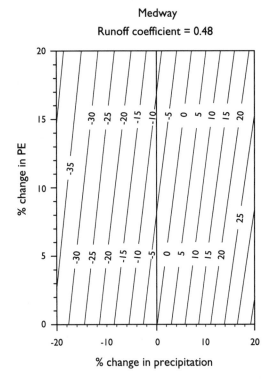

Medway
Runoff coefficient = 0.48

Scenario studies

Table 5.4 shows the components of the average annual water balance — precipitation, potential evaporation, actual evaporation, snowfall and runoff — together with the percentage change under the 1996 CCIRG scenario.

Changes in actual and potential evaporation

According to the 1996 CCIRG scenario, annual potential evaporation increases by up to 48% in southern Britain but can decrease in the north west of Scotland: none of the study catchments, however, had a decrease, and the smallest increase is just over 4%. The change in *actual* evaporation depends also on the assumed change in rainfall and varies considerably between catchments. Figure 5.3 shows the percentage change in actual and potential evaporation for each catchment, plotted against the runoff coefficient.

There are two main features to note from both Figure 5.3 and Table 5.4. First, the increase in actual evaporation (AE) is generally *less*, in both percentage and absolute terms, than the increase in PE. Second, although this is less clear because the rainfall change varies between catchments, the difference between the percentage change in actual and potential

TABLE 5.4 Percentage change in annual precipitation, potential evaporation, actual evaporation , snowfall and runoff under the 1996 CCRIG scenario

	Precipitation			Potential evaporation			Actual evaporation			Snowfall			Runoff		
	A	B	C	A	B	C	A	B	C	A	B	C	A	B	C
Almond	895	975	8.9	478	511	6.8	422	447	6.1	62	40	-35.2	474	527	11.4
Lyne Water	957	1043	8.9	478	511	6.8	456	485	6.5	64	41	-35.9	502	557	11.1
Bedburn Beck	866	945	9.1	495	560	13.8	364	400	9.8	101	66	-35.1	502	545	8.6
Greta	1123	1227	9.2	505	573	13.6	433	477	10.2	153	104	-32.0	690	749	8.6
Dove	1031	1098	6.5	556	700	25.8	427	498	16.7	0	0	0.0	604	600	-0.7
Harpers Brook	619	646	4.5	561	722	28.9	434	483	11.5	0	0	0.0	185	163	-12.0
Wensum	673	717	6.5	581	816	40.4	414	498	20.5	0	0	0.0	259	219	-15.7
Colne	564	588	4.3	586	865	47.7	423	498	17.7	0	0	0.0	141	91	-35.7
Turkey Brook	659	686	4.0	565	782	38.4	473	530	12.1	0	0	0.0	187	155	-16.6
Thames	726	754	3.9	566	750	32.5	449	510	13.6	0	0	0.0	277	244	-11.9
Lambourn	730	753	3.2	565	800	41.6	496	592	19.4	0	0	0.0	234	162	-31.1
Medway	848	882	4.0	543	789	45.2	458	558	21.9	0	0	0.0	390	324	-16.9
Avon	753	779	3.4	587	808	37.6	400	478	19.4	0	0	0.0	352	301	-14.7
Tamar	1216	1264	3.9	572	808	41.3	533	674	26.4	0	0	0.0	683	590	-13.6
Teme	836	881	5.4	549	705	28.5	482	554	15.1	0	0	0.0	355	327	-7.9
Cynon	1790	1893	5.7	534	691	29.4	497	611	23.1	79	46	-41.5	1294	1282	-0.9
Ewenny	1363	1421	4.2	575	768	33.5	469	573	22.2	45	21	-54.3	894	847	-5.2
Conwy	2200	2383	8.3	514	591	15.1	468	529	13.1	232	158	-31.6	1732	1854	7.0
Eden	1156	1267	9.6	466	530	13.7	418	463	10.8	187	118	-36.6	738	804	8.9
Nith	1571	1721	9.5	445	464	4.1	405	420	3.6	166	115	-30.4	1166	1300	11.6

A = current climate (mm), B = 1996 CCIRG scenario (mm), C = percentage change

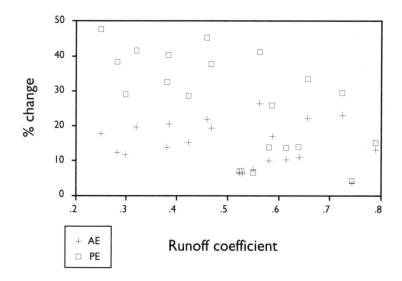

Runoff coefficient

FIGURE 5.3 Change in potential and actual evaporation

evaporation varies with runoff coefficient, with the greatest difference in catchments with the lowest runoff coefficient. Actual evaporation tends to increase by a smaller amount than potential evaporation because the water available for evaporation is limited. This is particularly true in "drier" catchments, so here the difference between the change in actual and potential evaporation is greatest. In wetter catchments, the actual rate of evaporation is close to the potential rate, so changes in the two are more similar. Four of the catchments show very little difference in the percentage change in actual and potential evaporation, and in one of these catchments the relative change in actual evaporation is in fact slightly higher. These four catchments have the lowest change in potential evaporation, and it appears that the greater the increase in potential evaporation, the greater the difference between the change in actual and potential evaporation.

Change in average annual snowfall

The snow model used in the study is crude and the simulated snowfall has not been validated, but some conclusions can be drawn. The model shows a major reduction in the amount of precipitation falling as snow: reductions range from 30% to 54% (Table 5.4). The effects of such a change are most clearly seen on monthly runoff totals.

Change in catchment average annual runoff

Figure 5.4 shows the percentage change in average annual runoff for each catchment under the 1996 CCIRG scenario together with, for comparative purposes, the changes under four scenarios based on the 1991 CCIRG report (Arnell and Reynard, 1993; 1996). The vertical bar for each catchment essentially defines the range in change under the 1991 scenarios, with the 1991 "best" estimates denoted by the thick bar. The catchments are ordered starting from north east (left of graph), moving south through eastern England to the south coast (centre of graph), and the running north up to west coast (right of graph).

There are two main points to draw from Figure 5.4. First, under the 1996 CCIRG scenario, annual runoff increases by up to 10% in the north and decreases by up to 35% in the Colne catchment in the south. Second, the change in annual runoff under the 1996 CCIRG scenario is generally outside the range of "best" estimates derived from the 1991 CCIRG scenarios, but is always within the extremes.

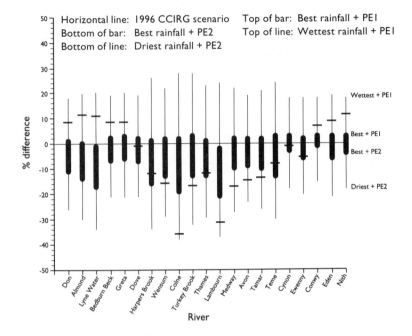

FIGURE 5.4 Change in average annual runoff by catchment: 1996 and 1991 CCIRG scenarios

Changes in average annual runoff across Britain

The water balance model was applied across the whole of Britain assuming that each $0.5° \times 0.5°$ cell represented a catchment. Figure 5.5 shows the percentage change in average annual runoff by the 2050s for each cell. Although the detailed pattern of change reflects to a certain extent the assumption of constant model parameters across the entire region, the general patterns are likely to be quite representative (and are consistent with the catchment-scale changes outlined above).

North of a line joining Liverpool, Leeds and York, average annual runoff is simulated to increase by over 5% by the 2050s, with increases of over 15% in some areas. South of a line connecting Hull with Bristol average annual runoff would decrease by over 5%, with decreases of more than 25% in south east England. Between these two regions, average annual runoff would vary by less than 5%.

VARIATION IN RUNOFF THROUGH THE YEAR

Changes in monthly flow regimes

Only in very specific circumstances will a change in climate have no effect on the distribution of flow through the year.

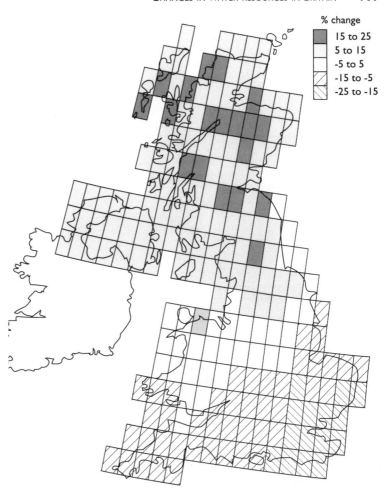

% change

■	15 to 25
□	5 to 15
□	-5 to 5
▨	-15 to -5
▨	-25 to -15

Figure 5.5 Percentage change in average annual runoff for the 2050s, by 0.5° × 0.5° cell.

Virtually all studies have found some seasonal variation in effect, with the changes reflecting not just the details of the climate change scenario but also catchment physical properties.

The most widely reported effect on monthly flow regimes is due to changes in the amount of precipitation that falls as snow. Many studies (e.g. Gleick, 1987; Lettenmaier and Gan, 1990; Mimikou and Kouvopoulos, 1991; Bultot *et al.*, 1992; Saelthun *et al.*, 1990) have found that global warming would lead to a reduction in both the spring snowmelt peak and the proportion of runoff occurring in spring in environments where the bulk of winter precipitation falls as snow. The proportion of flow occurring in winter generally increases, although whether the *absolute* magnitude increases depends on whether winter precipitation increases or decreases. Figure 5.6 illustrates this effect for two catchments in the headwaters of the Sacramento — San Joaquin basin in California (Lettenmaier and Gan, 1990), and two catchments in Norway (Saelthun *et al.*, 1990). The

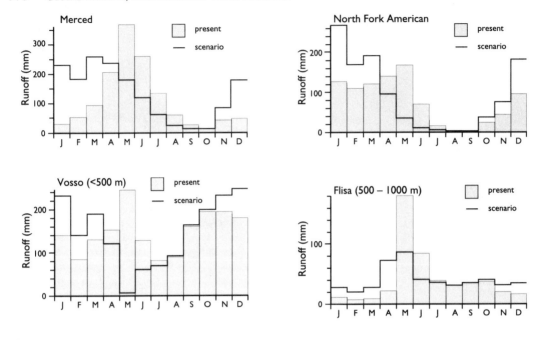

FIGURE 5.6 Monthly runoff in the Merced and North Fork American rivers, California (Lettenmaier and Gan, 1990), and the Vosso and Flisa rivers, Norway (Saelthun *et al.*, 1990)

Merced River is at a high elevation, so continues to show a spring snowmelt peak; however, this peak is supplemented by a winter rainfall peak. The North Fork American River has a mixed rainfall and snowmelt runoff regime at present, and under the scenario shown would change to one dominated by rainfall. Lettenmaier and Gan (1990) showed that the change in seasonal distribution of flows was caused largely by the rise in temperature, and hence the change from snowfall to rainfall, rather than a change in the seasonal distribution of precipitation. In the Norwegian study precipitation was assumed to increase throughout the year, with the largest percentage change in spring and summer. The effect of the rise in temperature, however, dominates the change in seasonal regime.

Figure 5.7 shows the monthly runoff for five Belgian catchments (Bultot *et al.*, 1988a; Gellens, 1993) and a catchment in Switzerland (Bultot *et al.*, 1992), under current conditions and a particular scenario. The scenario assumes an increase in winter rainfall, a decrease in summer rainfall, and an increase in potential evaporation throughout the year, and is similar to the 1996 CCIRG scenario for southern Britain. The Swiss catchment has a small snowmelt component and the virtual elimination of this under the change scenario is shown by the reduction in March runoff. The difference in change between the catchments largely reflects catchment geology. The Semois is a relatively impervious catchment in the Ardennes that experiences large increases in winter runoff and large reductions in summer runoff. At the other extreme, both the Dyle and the Aa are lowland catchments with important sand aquifers. Here, the extra rainfall received in winter recharges the aquifer and discharges slowly into the river through the year; summer flows are therefore maintained and sometimes increased. The Berwinne catchment is underlain by a chalk aquifer. Winter recharge is increased, but transmission times are short so this additional recharge reaches the stream relatively quickly and does not benefit summer flows.

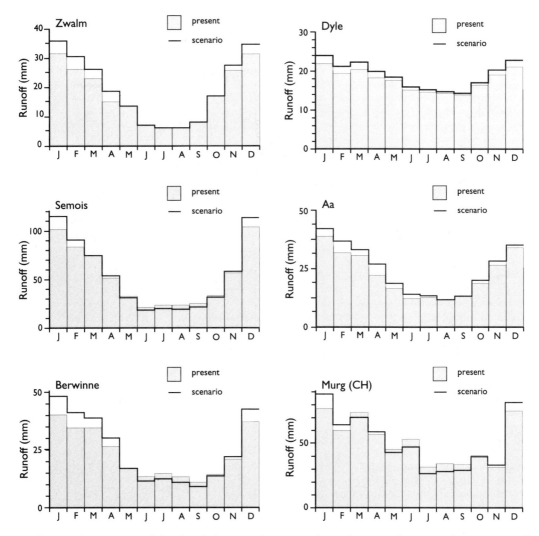

FIGURE 5.7 Monthly mean runoff for five Belgian catchments and one Swiss catchment, under current and changed climate (Bultot et al., 1988; 1992; Gellens, 1993)

Great Britain

Figure 5.8 shows mean monthly flow in six of the British case study catchments, for the baseline and 1996 CCIRG scenario. In the catchments in southern Britain (Harpers Brook, Medway and Tamar), there is a flow decrease in each month, with the greatest percentage decreases during summer: flows fall by more than 50%. Further north (e.g. the Greta catchment) there is an increase in flow during the winter, but little change during summer. Across Britain there is an increase in the flow variation through the year, with a greater concentration of annual runoff during winter. The effects of the reduction in snowfall can be seen in the Greta and Nith catchments, where, despite the increased precipitation, there is a reduction in spring (April — May) flows and a more substantial increase in winter (November — January) flows.

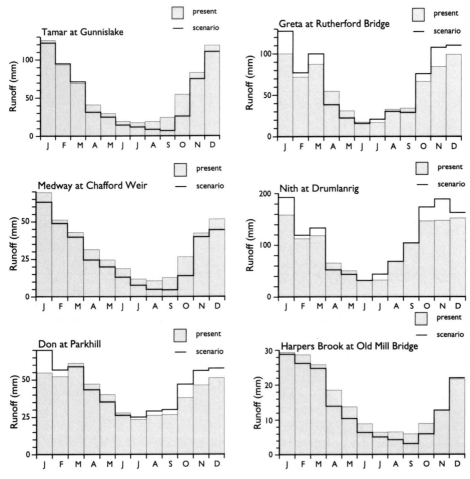

FIGURE 5.8 Monthly mean runoff for six British catchments under the baseline climate and the 1996 CCIRG scenario

There is relatively little difference in the effect of a given climate change scenario between lowland catchments with and without major aquifers, although there is a tendency for the monthly changes in the catchments with an underlying chalk aquifer to be proportionately smaller than changes in more responsive catchments. Some of any additional rainfall therefore contributes to groundwater recharge and discharges to the river later in the year, but the effect is not enough to sustain flows during summer. This is similar to the Belgian chalk catchment (the Berwinne), but differs from the results from the Belgian catchments with the deep sandy aquifer (the Dyle and the Aa), which have much lower transmissivities than the chalk. The conclusion is also different to that drawn in an earlier study with some of the same catchments, which used a monthly water balance model (Arnell *et al.*, 1990; Arnell, 1992); in this case the difference is probably due to the coarse time step of the monthly model and its less realistic simulation of runoff generation and routing processes.

FLOW FREQUENCIES

Although there have been many studies into the effect of climate change on monthly river flow regimes, only a few of these have considered changes in high and low flow extremes; these studies tend to be those using daily hydrological models. This section summarises the results of the Great Britain case study, drawing on information from the few other published investigations.

Changes in flow duration curves

The flow duration curve shows the proportion of time that river flow exceeds a given discharge value. The slope and shape of the flow duration curve characterise the variability in flows from day-to-day and largely define catchment low flow properties. If the flows are standardised by dividing by the mean, a steep flow duration curve characterises a catchment with very variable flows, whilst a shallow flow duration curve is representative of a catchment with very stable flows.

Figure 5.9 shows flow duration curves for three of the Great Britain study catchments: Harper's Brook, the Tamar and the Don, under the baseline climate and the 1996 CCIRG scenario. In the two southern catchments the future flow duration curve lies below the present curve, indicating that a given magnitude of flow would be exceeded *less* frequently in the future. For example, the flow currently exceeded 80% of the time in the Harper's Brook (0.12 m^3 s^{-1}) would only be exceeded 70% of the time under the 1996 CCIRG scenario; in the Tamar, the same frequency discharge would in the future be exceeded just 60% of the time. The steeper curves under the change scenario indicate the greater variability in flow through the year. Flow duration curves for catchments in northern Britain (such as the Don) are similar under the baseline and changed climates, with the greatest differences (in terms of frequencies) for flows above the median.

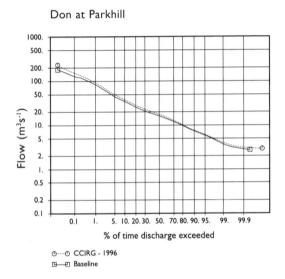

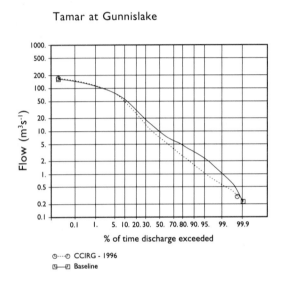

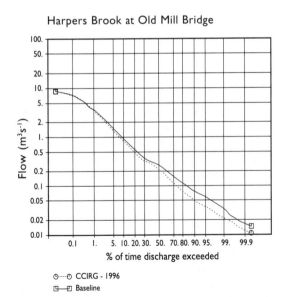

FIGURE 5.9 Flow duration curves for three of the British study catchments: baseline and CCIRG scenario

Changes in low flow frequencies

At first, it would appear obvious that a reduction in rainfall would lead to an increase in the rate of occurrence of low flow extremes. However, this will not necessarily occur in practice because low flows in a catchment are influenced by antecedent conditions, dating perhaps over several months.

Table 5.5 shows changes in the frequency of low discharges in five Belgian catchments (Gellens, 1991; 1993). The threshold frequency varied between catchments, varying from the flow exceeded 81% of the time to that exceeded 89% of the time. In three of the catchments (the Zwalm, Semois and Berwinne) the low flow value would be experienced slightly more frequently in the future and in the remaining two groundwater-dominated catchments the frequency of occurrence of low flow spells would reduce. There would be little change in the average duration of spells below the critical thresholds. Bultot *et al.* (1992) looked at low flows in Switzerland in a slightly different way; in the Murg catchment, the flow exceeded 82% of the time would fall by 10% (from 0.28 m^3s^{-1} to 0.25 m^3s^{-1})

TABLE 5.5   Frequency of occurrence of flows above defined low flow thresholds, and mean low flow spell duration: five Belgian catchments (Gellens, 1991; 1993)

	Current climate		Change scenario	
Zwalm	87.4	5.6	88.0	5.1
Semois	89.0	16.4	86.2	16.9
Berwinne	87.0	9.4	85.1	9.7
Dyle	81.2	8.4	84.7	8.3
Aa	86.1	8.2	88.7	8.2

Figure 5.10 shows the percentage change in the flow exceeded 95% of the time (Q95) in the Great Britain study catchments, under the 1996 CCIRG scenario: this frequency discharge is widely used in the water industry in Britain for setting licences and discharge consents. As in Figure 5.4, the catchments are ordered so that the left side of the diagram represents the north east, the centre represents the south and the catchments on the right are in the north west. In the northern catchments there is less than 5% change in Q95, but in the southern catchments the change is generally greater than the percentage change in summer *monthly* flows. This reflects the reduction in summer water availability, but the details of the change are also influenced by catchment geological conditions.

Low flows in Britain tend to occur in late summer after prolonged dry periods. They depend on:

- the interaction between the recession characteristics of the catchment (the rate at which flows decline following the end of rainfall),

- the amount of effective rainfall before the dry period, and
- the duration of the recession period before the next input of effective rainfall.

The first of these characteristics is largely influenced by basin geology and soil properties, whilst the second two are climatically determined. The effect of climate change on low flow properties in a given catchment therefore depends on the way in which changes in climate characteristics (effective rainfall amounts, durations of dry spells) combine with catchment physical properties. Low flows in highly responsive, high evaporation catchments would be expected to be little affected by changes in climate, because low flows are already low. At the extreme, catchments with a current low flow of zero (very rare in Britain) would not show any change at all. This is seen to an extent in the Turkey Brook catchment, which shows a much smaller change in Q95 than other catchments in southern Britain. Low flows in responsive upland catchments, however, would be more affected by a summer deficit. In these catchments summer rainfall is currently able to contribute to streamflow because potential evaporation is relatively low. Any increase in potential evaporation and decrease in rainfall would have a major effect on water availability and reduce summer flows: however, under the 1996 CCIRG scenario such catchments in northern Britain would experience an increase in summer rainfall.

In groundwater-dominated catchments, low flows are dependent on the amount of recharge over winter, so reflect the integration of changed climatic inputs over a period of several months; change in total winter rainfall is therefore important in determining changes in summer low flows. Summer flows on the Wiltshire Avon, for example, change by a smaller percentage than in other southern catchments (Figure 5.10), although the chalk-fed Lambourn shows a similar change to other catchments.

Changes in high flow frequencies

Popular accounts of global warming frequently suggest an increase in the occurrence of floods and flood disasters as a result of the enhanced greenhouse effect. However, there have been very few studies addressing the issue directly, due largely to the difficulties in defining credible scenarios for changes in flood-producing climatic events (Beran and Arnell, 1996). It is important to distinguish between rain-generated floods and floods associated with snowmelt (which may often be enhanced or triggered by rainfall).

The amount of rainfall that is required to generate a flood depends on catchment physical characteristics and the state of the catchment when the rain is falling, for example, less rain will be required to generate a flood if the catchment is already wet. The physical properties affecting the translation of rainfall to flood include gradients, catchment soil characteristics and storages within the catchment, such as lakes or wetlands. Small catchments with impermeable soils will be very sensitive to changes in short-duration intense rainfall. Larger catchments, or those with more permeable soils, will be sensitive to changes in catchment wetness and longer-duration rainfall totals.

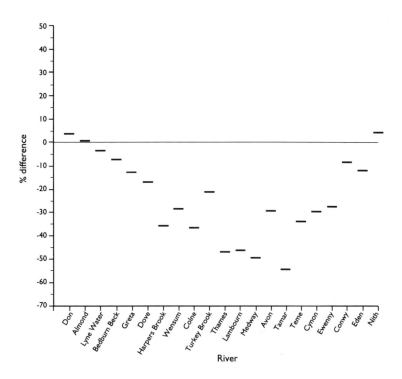

FIGURE 5.10 Percentage change in 5 under the 1996 CCIRG change scenario

Global warming can be expected to produce changes in the frequency of intense rainfall in a catchment for two main reasons. First, there may be a change in the path and intensities of depressions and storms bringing rainfall to the catchment, and second, there will probably be an increase in convective activity (Whetton *et al.*, 1993). The magnitudes of snowmelt floods are determined by the volume of snow stored, the rate at which it melts, and the amount of rain that falls during the melt period. If the amount of precipitation falling as snow declines, then snowmelt flooding might become less frequent. However, the temperature effect might be outweighed by increased precipitation, perhaps during extreme storms. Also, winter precipitation falling as rain, rather than snow, might produce flooding soon after it falls, shifting the flood season earlier in time.

There have been a few attempts to quantify potential changes in flood magnitudes. Lettenmaier and Gan (1990), for example, simulated changes in four catchments within the Sacramento-San Joaquin basin in California. They found a general decrease in the number of snowmelt events, and both an increase in annual maximum flood magnitudes and a shift forward in the dates of annual maxima. The peaks were higher because precipitation was no longer stored in the snow-pack and reached the channel quickly. In Norway, Saelthun *et al.* (1990) simulated decreases in spring snowmelt floods, and increases in autumn floods (Table 5.6). The change in spring floods was particularly large in "marginal" catchments or elevation zones, where the spring snowmelt flood is neither totally dominant (where a rise in temperature would not be sufficient to eliminate it) nor relatively minor. In some catchments, such as the upland Flisa, this meant a reduction in the annual maximum flood; in others, such as the Vosso, this resulted in an increase in the annual maximum. The coefficient of variation (CV: standard deviation divided by the mean) of annual flood magnitudes also changed, although it is difficult to see a consistent pattern. The CV generally increased, although the magnitude of increase varies.

		spring mean	CV	autumn mean	CV	annual max. mean	CV
Vosso:	1000-1500 m	-25	15	29	-13	-5	20
	500-1000 m	-2	104	29	-2	25	13
	<500 m	20	40	17	3	21	25
Flisa:	1000-1500 m	-21	21	23	24	-19	14
	500-1000 m	-43	30	27	-3	-38	24
	<500 m	-32	42	41	-25	-22	37

TABLE 5.6 Percentage change in the mean and coefficient of variation of annual floods by season, for ttwo Norwegian catchments divided into three elevation bands (Southern et al. 1990)

Table 5.7 shows the change in the frequency with which given threshold peak discharges are exceeded in five Belgian catchments, under a change scenario assuming increased winter rainfall (Gellens, 1991; 1993). In all the catchments studied, discharge would be above a defined threshold (two or three times the mean flow) more frequently in the future. For most, flows would be above the defined thresholds for approximately five extra days per year. In the Murg catchment in Switzerland, there was little change in the flow exceeded on average 18 days per year, but the annual maximum peak discharge was simulated to increase by approximately 10% and the coefficient of variation of the annual maxima increased by 15% from 0.19 to 0.22 (this was based on a simulated sample of just ten years: Bultot *et al.*, 1992).

TABLE 5.7 Change in frequency of occurrence of flows above defined thresholds per annum: Belgian catchments (Gellens, 1991; 1993)

| | Threshold value =2 × current mean flow | | Threshold value = 3 × current mean flow | |
	Current no. days > threshold	Future no. days > threshold	Current no. days > threshold	Future no. days> threshold
Zwalm	49.8	56.7	25.0	32.1
Semois	52.5	58.0	22.3	27.3
Berwinne	48.2	56.7	23.2	29.2
Dyle	10.2	15.2	1.2	2.1
Aa	43.4	52.2	14.6	19.7

At a larger spatial scale, Kwadijk and Middelkoop (1994) investigated changes in flood risk along the River Rhine, using a monthly simulation model and empirical relationships between monthly and peak values. They found large changes in the frequency of given threshold events and showed that an increase in precipitation and an increase in temperature could lead to major increases in both flood frequencies and the risk of inundation.

The hydrological model used in the Great Britain case study was not designed to simulate peak flows reliably and, as shown in Chapter 4, extreme high flows were generally poorly simulated. However, some indications of possible changes in the occurrence of high flows can be obtained by examining relatively frequent high flow events, which are simulated much more accurately.

Figure 5.11 shows the percentage change in the magnitude of Q5, the discharge exceeded 5% of the time (i.e. 18 days per year on average), under the 1996 CCIRG scenario, for each study catchment. In northern Britain (the left and right of the diagram), Q5 increases by between 10 and 25%. In southern Britain, there is a general decrease, but with considerable variability. On the Thames, there would be no change in the magnitude of the flow exceeded 5% of the time, but in the Colne this flow would decrease by 50%. The proportion of time that the current Q5 is exceeded depends not only on the change in the magnitude of flows, but also on the slope of the flow duration curve. A small change in magnitudes in a catchment with a gently-sloping curve can produce very large changes in the frequency of exceedance, whilst a large change in a catchment with a steep curve could have little effect on occurrence. For example, the magnitude of Q5 falls by 25% in the Lambourn catchment, which has a very flat flow duration curve, but the frequency of the current Q5 falls from 18 days per year to an average of just 1.8.

Before leaving this dicsussion of the potential effects of climate change on flood occurrence in Britain, it is important to

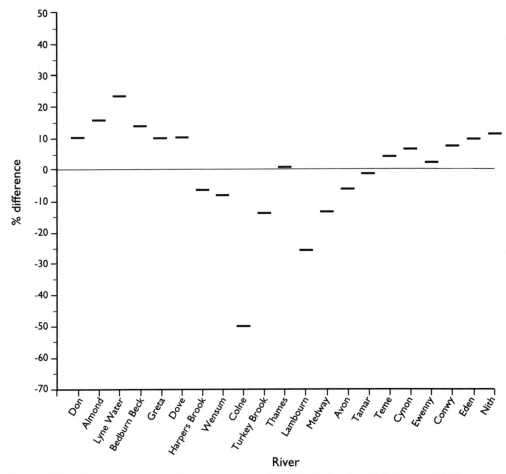

FIGURE 5.11   Percentage change in the flow exceeded 5% of the time Q5, in the British study catchments

emphasise a few limitations with the study. Although the calculations imply a reduction in high flow magnitudes in the south, this does not necessarily mean that high *floods* will be lower. The daily model used did not simulate high floods (the annual maximum peak discharge and higher) well, due both to its spatial resolution and to its coarse (in terms of flood generation) temporal resolution. The scenario used assumes a change in the magnitude of monthly rainfall and, by also assuming a change in the number of days on which rain falls, changes in average daily rainfall intensity can be inferred. However, the scenario does not explicitly cover changes in the frequency of extreme rainfall — that which causes floods — and it is quite possible that a small change in average daily rainfall intensity could hide a much more considerable, and significant, change in the frequency of occurrence of extreme rainfalls.

CHANGES IN GROUNDWATER RECHARGE AND GROUNDWATER LEVELS

Groundwater is an important source of water in many countries. A third of the supplies in England and Wales are drawn from groundwater, and the proportion is considerably higher in parts of the south: in other countries and regions groundwater is the most important source of water. There have, however, been very few studies of the effects of climate change on groundwater recharge and water in aquifers. Ghassemi *et al.* (1991) inferred qualitative changes in recharge in different aquifers in Australia from knowledge of recharge processes and possible changes in rainfall and evaporation, but did not use a quantitative model. Vaccaro (1992), in the most comprehensive assessment so far of groundwater recharge, used a daily recharge model to estimate changes in recharge in the semi-arid Ellensburg basin, on the Columbia plateau in Washington, USA. He used a stochastic weather generator and a scenario based on the average of three general circulation models, and also considered effects under both the predevelopment land use (grassland, sagebrush and forest) and current land use (irrigated and un-irrigated land). Under the predevelopment land cover, recharge would increase by around 10%, both because precipitation increased under the scenario considered and because there was increased infiltration from the spring snowpack. With the current land use, recharge is nearly four times higher at present than under the predevelopment land cover largely because of the infiltration of irrigation water. Under the climate change scenario, recharge with the current land use would *reduce* by 40%, despite the increased rainfall. This is because more of the irrigation water is evaporated and also, to a lesser extent, because there would be less extra infiltration from the snowpack, which is less extensive under the current land use than with the predevelopment land cover. Incidentally, Vaccaro's (1992) study illustrates the importance of considering the effects of human intervention on the sensitivity of the natural landscape to change, as well as indicating the significance of assumptions about future water use (specifically irrigation applications) on the impacts of climate change: this point will be returned to in Chapter 8.

The major aquifers in Britain are the chalk and greensands, Triassic sandstones and Jurassic limestones (Figure 4.3), with shallow sand and gravel aquifers of local importance. Recharge tends to occur during winter, once the summer soil moisture deficits have been filled during autumn, and before deficits begin to build up again in spring. The volume of recharge is therefore dependent upon the amount of winter rainfall, and the duration of the recharge season, which is determined by autumn and spring rainfall and evaporation. With no change in the duration of the recharge season, an increase in winter rainfall would tend to increase recharge, although it is possible that a larger proportion of the extra rainfall would not fill groundwater stores but would instead run directly off into rivers (e.g. during the wet late winter of 1990, Marsh *et al.*, 1994). Changes in recharge to British aquifers, then, will depend on the extent to which the increased winter rainfall is offset by a shorter recharge season, and the amount of any extra winter rainfall which goes rapidly to streamflow.

Hewett *et al.* (1993) applied some scenarios developed by Arnell *et al.* (1990) to the chalk aquifer in north Kent, and simulated a general increase in recharge. The scenarios assumed an increase in winter rainfall and an increase in potential evaporation, but the rainfall increase was larger than assumed under the 1996 CCIRG scenario, and the increase in potential evaporation smaller. Cooper *et al.* (1995) calculated recharge under the 1991 CCIRG scenarios, using a daily water balance model (in fact, the model described in Chapter 4) applied to a chalk aquifer catchment in central southern England and a Triassic sandstone catchment in the English Midlands. Assuming the change in rainfall as specified by the 1991 CCIRG scenario (an increase in winter of 8%), the change in recharge was shown to depend on the assumed change in potential evaporation, and hence the duration of the recharge season.

The same analysis was repeated, using the same model and the same two catchments, for the current study using the 1996 CCIRG scenario. Table 5.8 shows the percentage change in recharge under the 1996 scenario and also, from Cooper *et al.* (1995), under the 1991 CCIRG scenarios. Under the scenarios considered, it appears that the effect of the shorter recharge season is likely to outweigh the effects of the increased winter rainfall, and that recharge could reduce.

	Chalk aquifer, central- southern England	Triassic sandstone, Midlands
1991 CCIRG rainfall change scenario, plus evaporation scenario PE1 (PE up 9%)	-2	2
1991 CCIRG rainfall change scenario, plus evaporation scenario PE2 (PE up 29-30%)	-21	-13
1996 CCIRG scenario	-30	-13

TABLE 5.8 Percentage change in average annual recharge: chalk aquifer in central southern England and Triassic sandstone catchment in the English Midlands. 1991 CCIRG scenario results from Cooper *et al.* (1995)

A change in recharge would have an effect on groundwater levels within the aquifer, with a consequent effect on river baseflow. Wilkinson and Cooper (1992) applied an idealised model of the aquifer-river system to hypothetical aquifers with parameters characteristic of chalk (an aquifer with high transmissivity) and Triassic sandstone (which has low transmissivity), and examined the evolution of the change in level following a change in recharge. In a later study (Cooper *et al.*, 1995), the approach was refined to incorporate spatial variability in transmissivity within the aquifer, a realistic pattern of recharge (rather than a synthetic sine wave), and the use of the 1991 CCIRG scenarios. In the Triassic sandstone aquifer, groundwater storage and baseflow to the river would be little affected under 1991 CCIRG scenario with a modest rise in evaporation, but both would decline consistently by around 12% under the high evaporation change. In the chalk aquifer, however, the modest rise in evaporation results in reductions in storage and baseflow to rivers in autumn, and the larger increase both delays the point of minimum storage by between two and four weeks and leads to a reduction in minimum storage of around 40%: baseflows are therefore also reduced. The chalk aquifer was found to respond much more rapidly to an evolving change, and storage generally kept in equilibrium with changes in recharge inputs. However, the low-transmissivity of the Triassic sandstone aquifer means that it cannot maintain equilibrium, and groundwater storage and baseflow in any one year are dependent on the accumulated change so far.

Finally, groundwater levels in coastal aquifers might be affected not only by changes in recharge, but also by changes in sea level, and hence base level. In principle, a rise in sea level would alter hydraulic gradients and possibly therefore also change groundwater flow patterns. Clark*et al.* (1990) identified the coastal aquifers at risk from a rise in sea level, but concluded that in general the amount of rise anticipated due to climate change would have a small and very localised effect on both the intrusion of salt water into aquifers, and hydraulic gradients.

CHANGE IN RUNOFF AND RECHARGE IN BRITAIN: A SUMMARY

The possible effects of global warming on river flow regimes have just been reviewed, looking at the annual and monthly scales as well as high and low flow extremes. It is clear that there is considerable uncertainty about the size, and in many cases even the direction, of possible change and this is largely due to differences between climate change scenarios. However, there are some general conclusions, emphasised by the results of the Great Britain case study.

The volume of runoff

Total runoff volume is more sensitive to a change in precipitation than to a change in temperature or potential evaporation, and this has been very widely reported. A given proportional change in precipitation has a greater proportional effect on runoff in drier areas (with a larger proportion of precipitation consumed in evaporation) than in wetter areas, and this differential amplification too has been frequently noted.

In the Great Britain study, it was found that both the percentage and the absolute change in potential evaporation were larger than the change in actual evaporation, because actual evaporation is limited by water available. The difference between the changes in potential and actual evaporation is greatest in drier catchments, where the runoff coefficient is low and evaporation is heavily constrained by water availability. Also in Britain, it was found that the increase in temperature which might arise by the 2050s under global warming would lead to a significant reduction in snowfall, and hence in upland catchments an even larger proportion of annual precipitation would fall as rainfall.

Under the 1996 CCIRG scenario, there would be an increase in average annual runoff in the north of Britain (of between 5 and 15%), and a decrease in the south (of between 5% and 15%, but up to 25% in the south-east). However, there is considerable uncertainty about these changes, and different scenarios (e.g. Arnell and Reynard, 1993, 1996) can give quite different changes. In terms of water resources (see chapter 8) a reduction in rainfall in winter and spring — or no change — would have the greatest significance.

The timing of runoff through the year

In snow-affected areas, changes in the timing of flow through the year are determined largely by changes in temperature, and hence changes in snowfall and snowmelt. In general, studies have shown a reduction in the spring snowmelt peak, and an increase in flows during winter. In humid temperate catchments, the effect of a given change scenario depends on the current climate of the catchment and catchment geology. For example, flows in a catchment with a slowly-responding aquifer may be sustained through a drier, warmer summer by the slow discharge of increased winter recharge; in such catchments, the change in flow through the year would depend largely on changes in winter recharge.

Rivers in Britain under nearly all scenarios would experience an increased seasonal variation in flow, with proportionately more of the total runoff occurring during winter. In southern Britain, reductions in summer flows could be very substantial. The major reduction in snowfall and snowmelt in the few British catchments with regimes affected by snow would lead to large reductions in spring flows and increases in winter runoff, and consequently alterations in the seasonal distribution of flow.

Groundwater recharge

There have been very few studies into potential effects on groundwater recharge. Indications are, however, that the effects of increased winter rainfall would be offset by a shorter recharge season, so that groundwater recharge would be reduced. This tentative conclusion needs to be further investigated, given the importance of groundwater for water supplies in Britain.

High and low flow extremes

There have been rather fewer studies into changes in high and low flow extremes than into possible changes in monthly runoff regimes.

A reduction in rainfall would often lead to reduced low flows, especially when coupled with higher evaporation, but the precise effects would depend on catchment geology, changes in the length of dry spells and the amount of rainfall before the dry spell begins. In catchments with slowly-responding aquifers or significant amounts of storage, low flows may not become more extreme, and may even increase, despite a decrease in seasonal rainfall.

In the British case study catchments, however, low flows were decreased under all the scenarios which included a fall in rainfall, and in no catchment studied were the effects of catchment storage enough to outweigh the effects of the reduced rainfall. The percentage change in low flows (specifically the flow exceeded 95% of the time) was greater than the percentage change in monthly flows.

An increased risk of flooding under the enhanced greenhouse effect figures large in popular accounts, but there have been very few catchment studies, due largely to the difficulties in defining credible scenarios for change in flood-producing precipitation. In snow-affected catchments the reduction in the snowmelt peak may be offset by larger increases in winter rainfall floods. In catchments dominated by rainfall, the effects of a change in rainfall amounts and intensities would depend on catchment properties, including catchment size, slope and amount of storage; a small catchment would be most sensitive to changes in short-term rainfall characteristics, whilst a larger catchment with a lot of storage would be more sensitive to changes in rainfall totals over a period of time.

In the Great Britain case study, high flows (specifically the flow exceeded 5% of the time) increased in northern catchments, and decreased in the south. However, this may not be a very good guide to changes in *flood* magnitudes, because floods are generated by intense rainfalls and these were neither well simulated nor explicitly assumed to change under the 1996 scenario.

This chapter has concentrated on changes in "average" hydrological characteristics by the middle of the 21st Century, assuming a stable climate. The next chapter considers transient changes in river flow regimes between 1990 and 2050.

CHAPTER 6

Changes in river flows over time

The description of the hydrological characteristics of a future world with a stable climate, warmer than the present climate, has given very useful information on the potential magnitude of the effects of global warming. In practice, however, an enhanced greenhouse effect would lead to a gradual change in climate: there would be no "stable" future warm world.

Although there have been very few studies, there are three reasons why it could be useful to investigate the evolution of hydrological characteristics over time (studying the *transient* effect of global warming). First, such studies would contrast a climatic trend with year-to-year variability, and give insights into the ability to detect climate change effects. Second, transient simulation studies give an indication of the potential rate of change, which might be very important in terms of adaptation to change. Finally, transient simulations could provide information on when certain critical thresholds are likely to be crossed.

This chapter presents some results of transient simulations from the Great Britain case study, backed up by research conducted elsewhere.

CREATING TRANSIENT SCENARIOS

Transient GCM experiments (Murphy and Mitchell, 1995) simulate the effect of a gradual increase in greenhouse gas concentrations, and could in principle be used to define transient scenarios for hydrological impact assessments. However (as noted in Chapter 5) it is difficult to use GCM output directly in catchment studies, because GCMs do not necessarily simulate local climate well and because they have a coarse spatial resolution. It would be possible to apply a series of annual changes, derived from a transient GCM, to a baseline climate, but this has not yet been attempted.

The few studies that have considered transient scenarios have simply added a linear trend to a baseline climate. The fundamental problem with this approach is the assumption that change will be linear: transient GCM simulations show that the climatic effect of an increasing concentration of greenhouse gases may be non-linear at the regional scale (Murphy and Mitchell, 1995). Wolock and Hornberger (1991), McCabe and Wolock (1991a;1991b) and Wolock et al. (1993) all generated synthetic samples, with lengths varying from 60 to 300 years, with an assumed linear trend. The current study applied a linear trend to an observed 30-year baseline, doubled in length to produce a 60-year transient series (Chapter 4). This procedure was adopted because accurate synthetic climate data could not be generated for all the study catchments. It has the disadvantage that the sequencing of years in the initial 30-year sample is duplicated, and then reproduced with the trend added.

HYDROLOGICAL CHARACTERISTICS IN AN EVOLVING CLIMATE

Figure 6.1 shows the change in rainfall, potential evaporation and river runoff for the Harpers Brook catchment in Northamptonshire, between 1990 and 2050, under the transient 1996 CCIRG scenario: Figure 6.2 shows the same for the Don catchment in north-east Scotland.

Clearly, any trend in annual and seasonal rainfall and runoff totals is small, relative to year-to-year variability, and this is true even under more extreme scenarios (Arnell and Reynard, 1996). Year-to-year variability in potential evaporation is smaller than variability in either rainfall or runoff, so the assumed trends in potential evaporation over the period 1990 to 2050 are easier to identify. The small size of any trend, relative to inter-annual variability, means that climate change may be very difficult to detect in river flow and rainfall data.

Detection of climate change

In Chapter 2, the difficulties of identifying a climatic trend in observed hydrological data were highlighted: how easy would it be to detect a trend that is known to be there?

Table 6.1 shows the occasions when a significant trend was identified in the transient simulations over the period 1990 to 2050, for the two catchments in Figures 6.1 and 6.2. Two tests of trend were used: i) a trend exists if the slope of the linear regression relationship between value and time is significantly different from zero; ii) a trend exists if Kendall's tau coefficient (Press et al., 1986) is significantly different from zero.

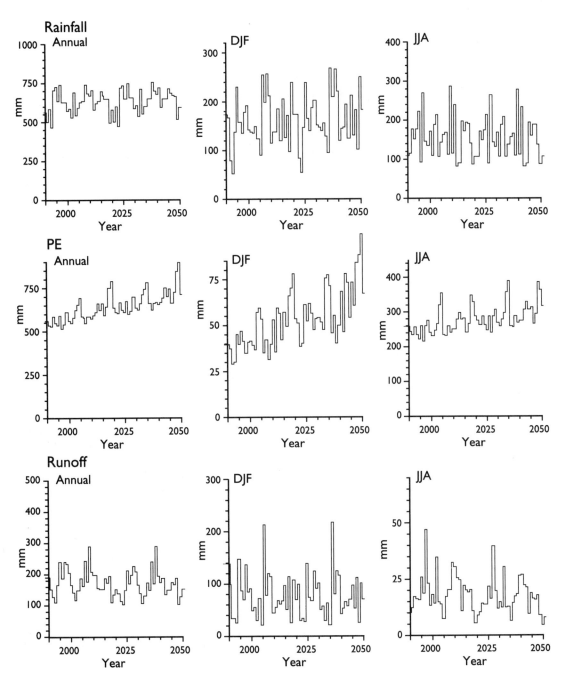

FIGURE 6.1 Seasonal rainfall, potential evaporation and runoff, under the transient 1996 CCIRG scenario. Harpers Brook at Old Mill Bridge

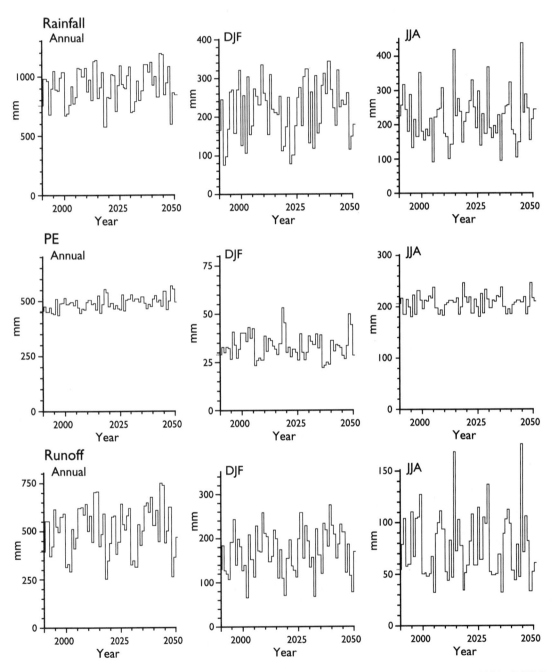

FIGURE 6.2 Seasonal rainfall, potential evaporation and runoff, under the transient 1996 CCIRG scenario. Don at Parkhill

Trends in rainfall are undetectable, using both tests, but trends in potential evaporation can be clearly identified, at least in the Harpers Brook catchment. No trends in runoff are detected, even though Figures 6.1 and 6.2 do give a visual impression of a decline in summer runoff in both catchments.

TABLE 6.1 Significant trends in seasonal rainfall, potential evaporation and runoff, 1990 - 2050: linear regression test (LR) and Kendall's tau test (τ)

	Annual		DJF		MAM		JJA		SON	
	LR	τ	LR	τ	LR	τ	LR	τ	LR	τ
32003: Harpers Brook										
Rainfall										
PE	**	**	**	**	**	**	**	**	**	**
Runoff										
11001: Don										
Rainfall										
PE	**			*	*			**	**	
Runoff										

**	significantly different from zero at 1%
*	significantly different from zero at 5%

Wolock and Hornberger (1991) conducted a set of simulation experiments to assess the likelihood of detecting an imposed climatic trend. They simulated flows in the White Oak Run catchment in the Blue Ridge Mountains of Virginia using TOPMODEL, a variable source-area simulation model. Synthetic climatic time series with an underlying trend were simulated, and Kendall's tau test was used to assess trend and its significance. Table 6.2 shows the proportion of simulation runs that showed a statistically significant trend in annual runoff and annual maximum peak after 60 years, under several change scenarios. Even after 60 years, only a small proportion of each set of simulations showed a trend, implying that the chance of detecting a trend would be small. Trends in different directions tend to counteract each other, and trends in annual maximum peaks were harder to identify than trends in annual runoff, because the year-to-year variability was higher.

TABLE 6.2 Percentage of simulations showing a statistically significant trend after 60 years, White Oak Run, Virginia (Wolock and Hornberger, 1991)

	Annual runoff		Annual peak discharge	
	Positive	Negative	Positive	Negative
No change:	0	3	0	0
Increase in storm intensity (20% after 60 years):	25	0	10	0
Increase in temperature (6°C in 60 years):	0	25	0	8
Increase in stomatal resistance (40% after 60 years):	42	0	7	0
Change in all three:	13	0	10	0

McCabe and Wolock (1991a;b) investigated the detectability of any trend in the Thornthwaite moisture index, defined on the annual scale as

$$I = 100 \left(\frac{P}{PE} - 1 \right)$$

In one study (McCabe and Wolock, 1991a) they simulated 300-year series of temperature (with a 0.4°C/decade trend) and precipitation (with no trend) for each of the 344 American climate divisions, evaluating trend using Kendall's tau statistic. Their results showed that the greater the year-to-year variability in the index, the longer it would take to detect a trend. Figure 6.3 plots detection time (defined by the year by which 50% of simulations show a significant trend) against the signal-to-noise ratio (the ratio of magnitude of change resulting from a 4°C warming to the standard deviation of the moisture index) for all 344 climate divisions. Even where the signal was largest relative to noise it would take at least 50 years to detect the effects of a trend in temperature. McCabe and Wolock (1991a) highlighted some of the shortcomings of their study — no change in precipitation and a relatively simple index of moisture availability — but the result emphasises the difficulties in detecting climate change trends. A parallel increase in precipitation would lengthen detection times, while a reduction in precipitation would shorten them.

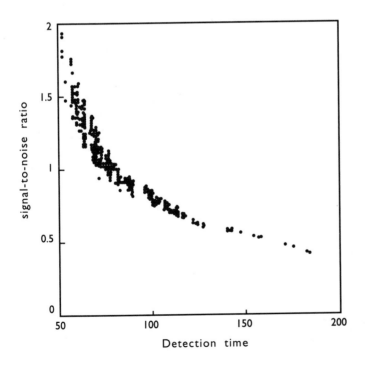

FIGURE 6.3 Relationship between the time needed to detect a trend in the Thornthwaite moisture index and the signal-to-noise ratio, by US climate division (McCabe and Wolock, 1991a)

In a subsequent study, McCabe and Wolock (1991b) explored the change in the Thornthwaite moisture index in the Delaware River basin, with the magnitude of transient change in temperature and precipitation determined from GCM scenarios. Table 6.3 shows the detection time (as defined above) at three locations in the Delaware basin, under three change scenarios. Clearly it would be many decades before a climate change could be detected using Kendall's tau statistic. The longest delay would be at Salisbury, the site with the greatest variability in precipitation from year to year.

TABLE 6.3   Number of years until the likelihood of detecting a trend in the Thornthwaite moisture index, as indicated by Kendall's tau, exceeds 50%. Three sites in the Delaware River Basin (McCabe and Wolock, 1991b)

Scenario	Salisbury	Philadelphia	Scranton
GISS	70-80	50-60	50-60
GFDL	60-70	50-60	50-60
OSU	>200	180-190	140-150

Changes in decadal means

While it may be difficult to detect changes at the annual scale, changes in longer-term averages may be clearer. Figure 6.4 shows the percentage difference between decade mean runoff under the 1996 CCIRG scenario and baseline mean runoff, by season, for the Harpers Brook catchment: Figure 6.5 shows the same for the Don. To a certain extent, the variation from decade to decade reflects the difference between a decade mean and a 30-year mean, but some patterns are apparent. In the Harpers Brook, decade mean runoff declines fairly consistently over the 60-year period, particularly during summer. In the Don, there is a gradual slight increase in annual runoff, but clearer increases in winter and autumn runoff.

Transient changes in the frequency of occurrence of low flow extremes

Figure 6.6 shows the number of days per year when flow in the Harpers Brook and Don catchments is less than the flow currently exceeded 95% of the time (Q95) under the 1996 CCIRG scenario. On average, flow should be less than Q95 on 18 days each year, although actually the frequency of such low flows varies considerably between years. Clearly, there is a consistent increase in the frequency of flows below Q95 in the Harpers Brook, and after 2025 flows would fall below this threshold nearly every year. There is little clear change in the frequency of occurrence of extreme low flows in the Don catchment.

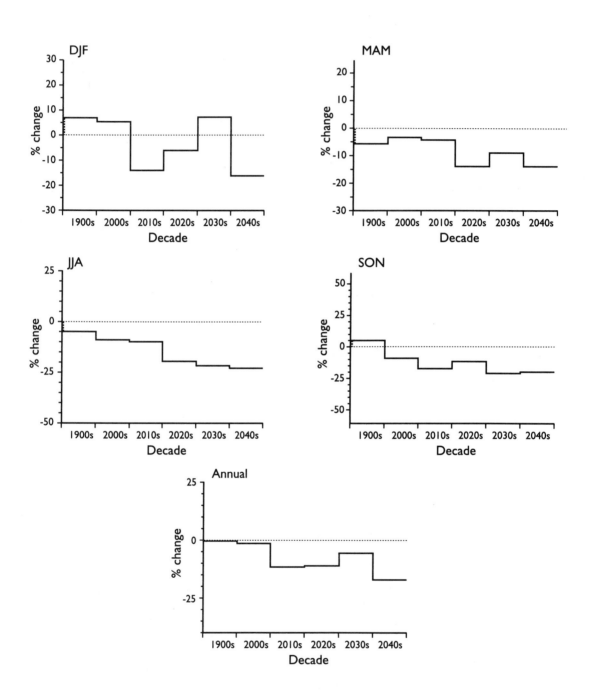

FIGURE 6.4   Percentage change in seasonal runoff by decade: Harpers Brook at Old Mill Bridge

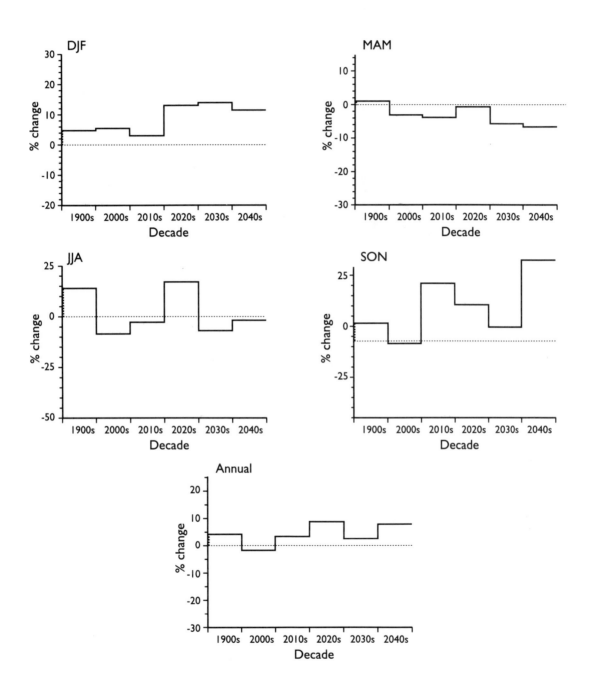

FIGURE 6.5   Percentage change in seasonal runoff by decade: Don at Parkhill

Figure 6.7 shows the Q95 flow in the Harpers Brook and the Don calculated for each decade. In the Harpers Brook, Q95 shows a consistent decline, but the pattern is less clear in the Don.

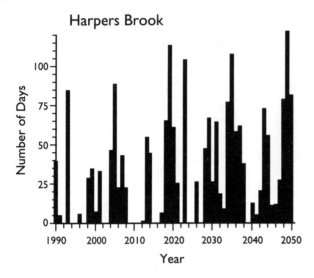

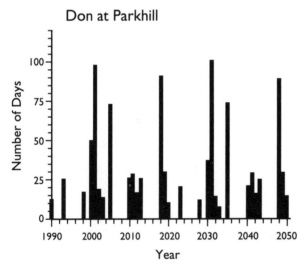

FIGURE 6.6   Number of days on which flow is less than the current Q95: Harpers Brook and Don catchments

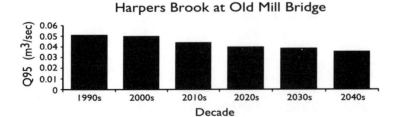

FIGURE 6.7A Q95 by decade: Harpers Brook catchment

FIGURE 6.7B Q95 by decade: Don
catchment

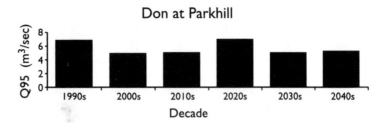

Don at Parkhill

CONCLUDING COMMENTS

Very few studies have looked at the pattern of change in hydrological regime as greenhouse gas concentrations increase, even though such inquiries could provide useful information on rates of change and, particularly, on the relationship between short-term variability and long-term trend.

The single most important conclusion from transient hydrological studies is that the effects of the imposed climate change trend on runoff are generally small relative to inter-annual variability. It may therefore be very difficult to identify a trend in flow data, even if one really does exist (see Chapter 2). Over the short-to-medium-term planning horizon (less than 20 years), year-to-year variability will dominate. Figure 6.8 shows five simulated 60-year time series, each with a substantial trend in mean and variance (Rogers, 1994): for the first 40 years at least,

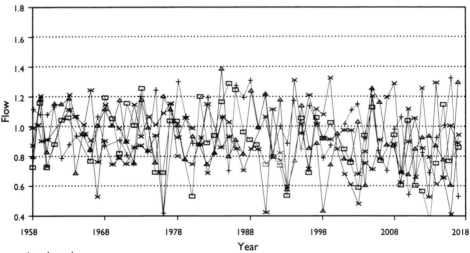

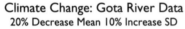

FIGURE 6.8 60-year simulated flow series with decreased mean and increased variance (Rogers, 1994)

year-to-year and between-sample variability are considerably more apparent than any climate change trend. The implications of this will be examined in the next chapter.

The year-to-year variability in hydrological behaviour is such that there may easily be periods of several years with flows significantly above or below some underlying progressive trend. At the decadal time scale a climate change trend would be clearer, although even here decade-to-decade variability masks some of the patterns. The hydrological signal is unclear because it is a result of the interaction between two inputs: precipitation and potential evaporation. Precipitation time series show considerable year-to-year variability, and climate change trends will be difficult to detect in these too. The climate change effect on potential evaporation should be significantly clearer. When potential evaporation increases and precipitation declines, runoff should show the greatest climate change signal. When an increase in evaporation is offset by a rise in precipitation, however, the signal may be very small. In some environments it is possible that a gradual change in precipitation and temperature could produce a non-linear change in runoff, where the hydrological regime changes from one type to another. An example would be when a snow-dominated regime changes to one in which snow ceases to be important every year. In a humid temperate environment like Britain, such non-linear changes are unlikely to occur.

Chapter 7

Changes in water quality

The term "water quality" describes the chemical characteristics of water as well as physical characteristics such as temperature, colour and suspended sediment concentration. Water quality in rivers and lakes depends on a complex system of inputs and feedback. The natural background streamwater chemistry is determined by atmospheric inputs, catchment geology and soil type, the processes by which water reaches the river network and by chemical and biological processes operating in the river. Superimposed on these natural controls are catchment land use, deposition of atmospheric pollutants, the discharge of used water from sewage treatment plants and other uses (such as industry and power generation), drainage from urban areas, and pollution incidents. The effects of climate change on natural" chemical processes will be superimposed on these anthropogenic factors, which are also varying through time for many reasons, including climate change.

In Britain, it is convenient to distinguish between upland and lowland systems (Jenkins*et al.*, 1993). Upland streams tend to be fast flowing and relatively oligotrophic, high in dissolved oxygen, and with catchments largely draining land which is not under intensive cultivation. Water quality in these systems tends to reflect the physiographic characteristics of the catchment, particularly bedrock geology and soil type. Lowland river systems, however, are generally slow flowing, with low turbulence, and therefore with dissolved oxygen contents below saturation. Such catchments also tend to be heavily used for agriculture so rivers have high nutrient concentrations and often drain large populations; rivers are therefore used as disposal routes for sewage and industrial effluents.

There are five ways in which climate change resulting from increasing concentration of greenhouse gases might affect water quality. A change in temperature will affect the rate of operation of biogeochemical processes that determine water quality. Changes in river flow volumes will affect dilution and residence times. Increased atmospheric CO_2 concentrations will affect the rate at which CO_2 is dissolved in water, and hence the rate of operation of many chemical processes. A change in flow pathways will alter the transport of chemical load from the land surface to the river network, and finally, a change in inputs of chemicals to the catchment will affect loads in the river. Before exploring these potential changes in more detail, it is necessary to summarise the effects of global warming on river water temperature.

CLIMATE CHANGE AND RIVER WATER TEMPERATURE

Temperature is perhaps the most important physical quality characteristic of river water. It not only affects biological and chemical processes in the river, but also influences aquatic ecosystems in general and the human use of river water. The rates of biological and chemical processes are temperature-dependent, and temperature influences the ability of water to absorb gases such as nitrogen, oxygen and carbon dioxide.

River water temperature in Britain is largely determined by air temperature, although there is a geological influence. Groundwater is generally cooler than direct runoff or runoff draining through soil, so catchments with a strong groundwater influence have lower water temperatures than catchments dominated by surface and near-surface sources. More locally, river water temperature is influenced by the degree of shading, so that the greater the amount of overhanging vegetation, the lower the water temperature. Human influences, specifically discharges from power stations and industrial plants, also affect water temperature. In south east England average water temperature can exceed 13°C, whilst in north west Scotland the average can be less than 8°C (Webb, 1992). The seasonal cycle has a range of between 10 and 12°C across Britain (with a smaller range in the west). Summer extreme maxima in the south-east can exceed 28°C, and although many rivers in the north experience minima below 0°C, rivers are very rarely ice-covered.

Changes in water temperature caused by global warming could be determined using theoretical thermal equilibrium models (as used by Cooter and Cooter (1990) in the southern United States and Stefan and Sinokrot (1993) in the north central United States), but changes in the meteorological inputs to these models are very uncertain. An alternative approach, used by Webb (1992) in Britain, is based on empirical relationships between air and water temperature, calibrated for each study site.

Webb (1992) found that the relationship between monthly mean air and water temperatures in Britain was generally very strong: on average, across 36 sites, variations in air temperature over time explained 92% of the variability in water temperature. In most cases, a given increase in air temperature would produce a slightly smaller increase in water temperature. Table 7.1 shows the sensitivity to change in air temperature in four different catchment conditions, taken from Webb (1992), together with the estimated change in water temperature for winter and summer in four locations in Britain, under the 1996 temperature change scenarios. In the "average" catchment, water temperature would rise by up to 1.6°C in summer and 1.8°C in winter in south east England, with lesser increases further to the north and west (increasing water temperature gradients across Britain). There was some evidence that summer water temperatures in small headwater streams would show a greater sensitivity to a given change in temperature than winter water temperatures, which would mean that the summer increases shown in Table 7.1 could be larger.

Webb (1992) also concluded from his regression relationships that there would be a greater increase in minimum temperatures than maximum temperatures, with a consequent reduction in the monthly range. The risk of high temperature extremes would be increased: in the Exe in Devon, a rise in average water temperature of 1.6°C would mean that the probability of temperatures greater than 25°C would rise from just 0.44% (1.6 days per year) to 1.07% (3.9 days per year). In contrast, the frequency of low water temperatures would decline.

TABLE 7.1 Sensitivity of summer and winter mean water temperature to change in air temperature (Webb, 1992), and increase (°C), by the 2050s, under the 1996 CCIRG scenario

	River type			
	Spring-fed /groundwater	Riparian shading	Average	Sensitive
Sensitivity*	0.5	0.7	0.9	1.2
Change in summer water temperature (°C):				
South-east England	0.9	1.3	1.6	2.2
South-west England	0.8	1.1	1.4	1.9
North England	0.7	1.0	1.3	1.7
North Scotland	0.6	0.8	1.1	1.4
Change in winter water temperature (°C):				
South-east England	1.0	1.4	1.8	2.4
South-west England	0.7	1.0	1.3	1.7
North England	0.6	0.8	1.1	1.4
North Scotland	0.4	0.6	0.7	1.0

* Sensitivity is change in water temperature per degree change in air temperature.

An increase in streamflow tends to reduce water temperature, and a decrease results in a rise in temperature, although the effect depends on the source of the extra water. Webb (1992) found that whilst variations over time in water temperature in the Exe catchment in south west England were largely due to variations in air temperature, a small (around 3%) but statistically significant amount of variability was related to changes in streamflow. In the main channel of the Exe, water temperature would rise by a further 0.5°C if flows were to halve — and, as shown in Chapter 5, such a change is possible. Summer water temperature could therefore rise by at least 2°C in some rivers in southern England under the 1996 CCIRG scenario. In small headwater streams, the effect of streamflow volume on water temperature is smaller than further downstream.

STREAM BIOCHEMISTRY

The chemical composition of river water depends basically on chemical inputs, chemical reactions and the time over which these reactions can take place.

Natural inputs to the river come from atmospheric deposition and the processes of weathering in the soil and rock. They depend on the soil and rock type, on flow pathways through the soil and rock, and particularly on the time taken for water to move to the river. "Unnatural" inputs include acid deposition, pesticide and fertiliser inputs to farmland (non-point sources) and point sources such as discharges from sewage treatment works, industrial works and pollution incidents.

Chemical reactions between species in water depend on factors including water temperature, acidity and concentrations of particular species and ions. In general, higher temperatures mean that the chemical and microbial processes which alter chemical composition operate more rapidly, although different processes are sensitive in different ways. Changes in the concentration of CO_2 in the atmosphere can also affect the chemical composition of water. The rate of solution of CO_2 in water is determined not only by water temperature, but also by the atmospheric concentration. Solution of CO_2 in water produces carbonic acid (H_2CO_3), which in turn may form the ions HCO_3^- and CO_3^{2-}. This process releases hydrogen ions H^+, which alters the acidity of the water: a change in atmospheric CO_2 concentrations may therefore alter stream acidity. Calcite ($CaCO_3$) occurs in many carbonate rocks (such as chalk and limestone) and dissolves in water. The chemical species produced depends on acidity, and the solubility of calcite depends on the concentration of HCO_3^- and H^+, which are themselves influenced by atmospheric CO_2 concentrations. A rise in CO_2 concentrations could therefore lead to a change in stream acidity and the rate of solution of calcite: however, the relative effect of changes in atmospheric composition and water temperature remain to be determined.

Changes in flow patterns will affect both the concentration of a chemical and may also affect residence times and hence the period over which processes can operate. In general, a reduction in flows would lead to increased concentration and increased time for processes to operate.

Nitrate concentrations

River and groundwater pollution by nitrates is a major concern in many developed countries. The nitrates derive from nitrogen both applied in agricultural fertilisers and released when grassland pasture is ploughed for arable crops. Nitrate concentration within a river reach depends upon the volume of inputs, the volume of flow and the rates of nitrification and denitrification. Both nitrification and denitrification are biologically-mediated reactions, and the amount of each is determined by residence time and water temperature. An increase in temperature would increase the rates of both nitrification and denitrification, but denitrification rates are most affected so, other things being equal, higher water temperatures would lead to a reduction in nitrate concentrations (Jenkins et al., 1993). Increased residence times, due to lower flows, would also result in lower nitrate concentrations because

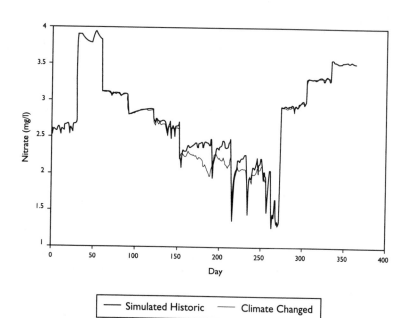

FIGURE 7.1 Nitrate concentrations in the Tamar under a climate change scenario (Jenkins et al., 1993)

denitrification could continue for longer. However, these effects might be offset by the reduced dilution due to lower flow volumes: the actual change in nitrate concentrations in a river will therefore depend on the temperature change, the change in streamflow volumes and the volume of nitrate inputs.

Jenkins *et al.* (1993) simulated nitrate concentrations in a number of rivers using the QUASAR quality simulation model and scenarios broadly similar to the 1996 CCIRG scenario. Figure 7.1 shows simulated nitrate concentrations on the Tamar in Cornwall (Jenkins *et al.*, 1993). Nitrate concentrations are reduced, particularly during summer, indicating that the effects of higher temperatures and increased concentrations outweigh the reduced diluting effect of the lower flows.

However, there are three other important potential effects of climate change on nitrate concentrations which are not accounted for in Figure 7.1. First, a change in agricultural practices triggered by climate change might lead to a change in inputs of agro-chemicals. Second, higher temperatures and drier soils would increase the rate of mineralisation of organic nitrogen in the soil, increasing the amount of nitrogen available to be washed into the river system. Third, peak nitrate concentrations occur in autumn when nitrates that have accumulated in the soil through prolonged summer dry spells are flushed into the river, as happened after the 1976 drought (Jenkins *et al.*, 1993). Autumnal flushing may increase if summers become drier.

Taken together, these three influences might lead to an increase in nitrate concentrations, and more particularly to an increase in peak concentrations. Immediately downstream of an input of effluent, the effect of reduced flows would be likely to outweigh the effect of the higher temperature, leading to increased nitrate concentrations locally (Jenkins *et al.*, 1993).

Dissolved oxygen and biochemical oxygen demand

The amount of oxygen in a stream — measured as a concentration of dissolved oxygen (DO) — determines the ecological health of the stream. It is controlled by four influences, namely the volume of water, the temperature of water, the consumption of oxygen during the decay of organic matter, and the amount of oxygen removed during the nitrification of ammonium: these last two together comprise the biochemical oxygen demand (BOD). The higher the water temperature, the lower the amount of oxygen that can be held in water, so a rise in temperature would lead to a reduction in DO. BOD is determined by the quantity and type of organic material in the river and, in many catchments, by the quantity and quality of point source pollutants. The amount of organic material in the river could alter as a result of changes in catchment vegetation

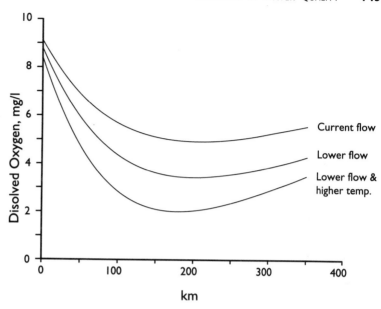

FIGURE 7.2 Effect of changes in flow and temperature on dissolved oxygen contents, downstream of a point source (Jacoby, 1990)

(particularly riparian vegetation) and the pathways through which water reaches the river.

A rise in water temperature would be expected to increase both the rate of reduction of oxygen contents (deoxidation) and the rate of recovery (reaeration), but the rate of deoxidation is more sensitive (Jacoby, 1990), so DO would fall. Figure 7.2 shows the effects of reduced dilution and increased water temperature on DO contents, at different distances downstream from a point source of pollution (Jacoby, 1990): the decrease in flow alone (by 25% in this example) leads to a reduction in DO, and the rise in deoxidation associated with a 2°C rise in temperature lowers DO further.

Jenkins *et al.* (1993) used the QUASAR water quality simulation model to simulate the effects of climate change on DO and BOD in a few British rivers. In the relatively clean Tamar (south-west England), higher temperatures in summer resulted in a decline in DO, but DO levels in winter were slightly increased (Figure 7.3a). In the Thames however, with a considerably greater BOD arising from effluent disposal, DO reduced throughout the year under all the change scenarios considered.

Algal blooms in rivers

Algal blooms can make a major contribution to BOD. They also affect the aesthetic quality of water, clog river intakes and lead to toxicity problems. The factors influencing their growth are complex, but it is likely that blooms are triggered by a combination of high temperatures and increased concentrations of nutrients, particularly phosphates.

(a)

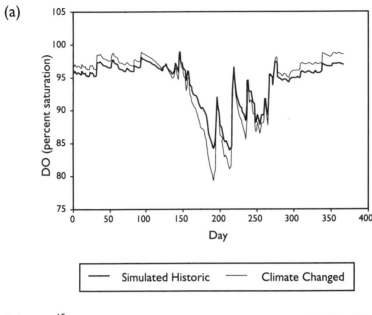

(b)

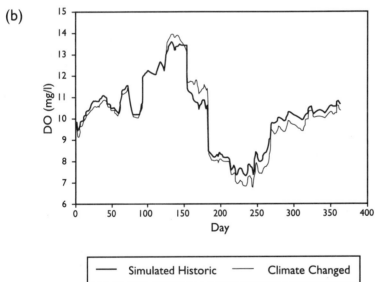

FIGURE 7.3   Effect of increased temperature and altered flows on dissolved oxygen concentrations: a) Tamar, b) Thames (Jenkins *et al.*, 1993)

Dense populations of algae can develop in isolated "dead zones" within pool and riffle sequences in rivers, and provide the seed for bloom outbreaks. Higher water temperatures will increase growth rates of algae and alter the pattern of succession of different species (Arnell *et al.*, 1994). The growth rates of blooms are also affected by flow volume and residence time: as flows fall and residence times increase, algal bloom growth is stimulated. Figure 7.4 shows the simulated concentrations of chlorophyll downstream of different initial input concentrations under a range of flows (Institute of Hydrology, 1990): with an

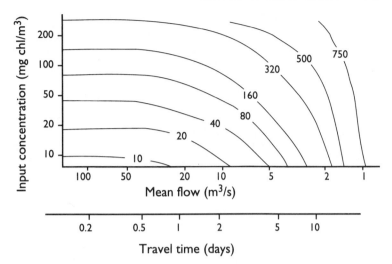

FIGURE 7.4 Simulated chlorophyll concentrations downstream of an initial inoculum: the River Thames at Maidenhead (Institute of Hydrology, 1990)

initial concentration of 50 mg m^{-3}, for example, downstream concentrations would be just under 50 mg m^{-3} with a discharge of 50 m^3 s^{-1}, 80 mg m^{-3} with a discharge of 10 m^3 s^{-1} and nearly 500 mg m^{-3} with a flow of 2 m^3 s^{-1}.

WATER QUALITY IN LAKES AND RESERVOIRS

Global warming would affect lake water quality by changes in throughputs and volumes, in water temperature and in the duration of ice cover (Jacoby, 1990).

Changes in loads of different chemical species brought in by rivers will affect the chemical balance of a lake or reservoir. A rise in temperature will affect thermal stratification and therefore the mixing of oxygen through the lake. In humid temperate environments such as Britain, lakes tend to stratify thermally in summer, with a warm, well-mixed oxygen rich epilimnion overlying a cool, oxygen-poor hypolimnion. As temperature falls during autumn, the stratification breaks down and water mixes. In lakes with a seasonal ice cover, mixing occurs in autumn, before the ice forms, and again in spring after the ice melts.

Increased temperatures are likely to increase the duration of stratification and, by changing rates of bacterial activity, will alter dissolved oxygen levels. For example, Blumberg and DiToro (1990) simulated a reduction in DO in Lake Erie of 1 mg l^{-1} in the upper layers and up to 2 mg l^{-1} in the lower layers, for a rise of between 3.5 and 4.2 degrees in air temperature. Stefan and Fang (1994) found even larger changes in a number of smaller lakes in Minnesota and simulated reductions in temperature at depth in some lake types. In many lakes, DO levels at the surface are close to saturation, and so would change relatively little as temperature rises. However, at greater depth oxygen content is related to lake size, trophic status and the duration of stratification. A rise in temperature would alter the dates when stratification begins and breaks down, thus lengthening the stratification season. In lakes with a seasonal ice cover, the duration of mixing and stratification is constrained by the period of open water. If ice cover were to be eliminated completely, lakes would mix continuously from autumn to spring, and stratify for longer in summer.

Wind too affects mixing and stratification, and hence water temperature and DO concentrations at depth. Deep British lakes, including most Scottish lochs and Lake District lakes, follow the standard pattern of temperate lakes, and tend to stratify towards the end of May and destratify during October: they rarely develop an ice cover. Surface temperature is related not just to the incoming solar radiation, but also to the intensity of turbulent mixing, which is a function of windspeed. The degree of stratification varies from year to year depending on incoming radiation and windspeeds. Changes in lake temperature in the future will therefore depend on changes in radiation (dependent on cloud cover) and changes in windspeed. Under the 1996 CCIRG scenario incoming radiation will increase in summer, but so will windspeeds (Chapter 4). The relative effect of these two changes on lake temperature has not yet been determined.

Shallow lakes such as Loch Leven, Scotland, do not develop thermal stratification: lake temperature closely follows the pattern of incident radiation (George, 1989). Given an increase in summer radiation, water temperature in such lakes is likely to rise and, following the results of the American studies cited above, dissolved oxygen concentrations will fall.

Reservoirs have temperature profiles and water quality similar to natural lakes. Jenkins *et al.* (1993) simulated nitrate concentrations in Farmoor pumped storage reservoir, Oxfordshire. They found that higher temperatures reduced nitrate concentrations because denitrification increases more rapidly than nitrification. However changes in residence time of water within the reservoir also affected nitrate concentrations, and the effects of changes in temperature were likely to be outweighed by changes in the loads of incoming river water.

Algal blooms have flourished on some British lakes and reservoirs in some recent years, due to a combination of higher temperatures, calmer weather and increased phosphate inputs. Higher winter temperatures and lower windspeeds during winter mean that more phytoplankton cells survive the winter, providing a larger base for an increase through summer, but the seasonal succession of phytoplankton is influenced by the timing and strength of mixing (George *et al.*, 1990). Bloom-forming species of blue-green algae, which cause the greatest problems to water managers, flourish during long, calm, warm spells. In deep lakes, such as Windermere, their summer growth is largely controlled by periodic mixing. Changes in the intensity of mixing, caused by variation in windspeed, alters the frequency of bloom growths. Under the 1996 CCIRG scenario, windspeed is generally assumed to increase, suggesting a decline in algal bloom problems on deep lakes, but more stable anticyclonic conditions during summer would lead to longer spells of calm weather and increased blooms. In some shallow, highly eutrophic lakes, dense blooms of blue-green algae form every summer, and they are less likely to be affected by climate change.

SEDIMENT, EROSION AND CHANNEL STABILITY

Changes in catchment vegetation, rainfall and hydrological regimes will affect erosion on hillslopes and in river channels, sediment transport and deposition, and river channel stability. However, there have been very few studies into the implications of global warming, and it may be difficult to separate climate change effects from the consequences of land use change.

Boardman *et al.* (1990) simulated the effects of changes in rainfall regimes on soil erosion in Britain, assuming no change in land use. Increased winter rainfall resulted in greater erosion from arable fields in lowland Britain, but lower losses in upland Britain because the warmer temperatures meant that ground cover lasted longer through winter. However, not all the sediment created by soil erosion reaches the river network, so implications for stream sediment yields are difficult to estimate. Higher peak flows would alter river channel erosion rates, but again effects on sediment yields are difficult to ascertain: there have been no studies. Channel stability too depends on the balance between river flows and sediment loads, and future changes are uncertain. An increase in winter flows is likely to *increase* channel instability, but there may be important thresholds in operation and changes may only occur once some critical value is exceeded. For example, a channel close to the boundary between braided and meandering would be more sensitive to change than one which has a clearly defined meandering channel.

Although there is much observational evidence of the effects of past climatic variability on channel stability, it is very difficult to estimate future changes because changes in the critical discharge values are uncertain and because there are very few process-based models which can predict channel change from altered inputs (Newson & Lewin, 1991).

OVERVIEW

The effects of global warming on stream and lake water quality are much less well understood than effects on water quantity, but it is clear that changes will depend very much on local climatic, geological and hydrological conditions, as well as on the environmental control measures in place. There appears to be a general deterioration in water quality with higher water temperatures, and particularly with lower river flows, although this will not be evident everywhere and is most likely in lakes and rivers that already receive high effluent inputs. There is a need to improve process-based models of river and lake water quality, in order to estimate better the effects of changes in climate.

Water quality is, of course, influenced by many factors other than climate change. Most important are agricultural practices (particularly the use of chemicals) and the degree of treatment of effluent discharges. Many changes in water quality will be determined more by legislative and economic pressures — such as water quality targets for individual river reaches, or the price of licences to discharge effluent — than by climate change.

CHAPTER 8

Implications for water resources and water management

The preceding chapters have examined the potential physical effects of global warming on river flows, groundwater recharge and water quality. This chapter considers the possible impacts of these effects on water users and the management of water resources. As indicated in Chapter 3, a large effect might have a small impact in some circumstances, while in others a small effect could have a very large impact. The relationship between effect and impact is not linear. Like the preceding ones, this chapter focuses on Britain but also draws on work from elsewhere.

ASSESSMENT OF IMPACT

The translation of physical effect into impact (with some implied value) is difficult for several reasons. First, and most important, water management is an adaptive process. Water managers (whether executives of major utilities supplying millions of customers or individuals responsible for their own supplies) will not sit back and suffer climate change effects without some form of incremental response. A water resource impact study which ignores water managers' operational response to change represents a worst case. The "impact" of climate change will be the sum of the economic, environmental and social costs involved in adaptation, and the costs of any residual consequences which cannot be mitigated. Adaptation will be based on information and experience available at the time, so will not be perfect. Different water management systems and different users, will be able to adapt to change in different ways. Some systems will be able to react quickly and effectively; others with longer planning horizons and lead times will respond more slowly, perhaps at greater cost.

Second, water managers face many changes and pressures

over the next few years and decades. At the global scale, the most obvious is population growth and the associated increased demand for water. Demands for water will become increasingly concentrated in space as urbanisation continues. Economic development tends to increase the volume of water demanded while technological changes in industry and agriculture tend to increase efficiency of water use. Technological developments in the water sector can stimulate change in water management; examples include the use of highly-integrated water supply systems, increased use of demand management rather than supply provision techniques (e.g. via water tariff structures) and the increasing cost-effectiveness of salt water desalination. Legislative provisions can be expected to change over the next few years. The water industry in England and Wales is still adjusting to privatisation in 1989 and has been forced to spend considerable sums on ensuring water meets quality standards set by the European Commission.

Finally, priorities given to different water uses and aspects of the water environment change over time. In many developed countries, water management over the past decade has paid much more attention to environmental impacts and consequences, and this trend is likely to continue. In part this reflects legislative changes but it also follows public opinion. The issues that determine the variability of "impact" for a given climate change will be further explored later in the chapter.

There are many users of water and several classifications of use. Two important distinctions are between *instream* and *offstream* uses, and between *consumptive* and *non-consumptive* uses. Instream uses exploit water within the river channel or lake: water does not need to be extracted. Offstream uses must take the water away from the river. Domestic, industrial and agricultural uses are offstream; instream uses include navigation, hydropower generation and aquatic ecosystems. Consumptive uses remove water from the river system, generally through evaporation: the best examples are irrigation and the use of water for cooling by industry and power generators. Most other uses are non-consumptive, although water abstracted from one part of the river system may be returned, after use, to another part, and may have a different chemical quality or temperature.

A change in hydrological characteristics may affect not only the use of water resources, but also the threats to economic and social well-being posed by water-related hazards. Specifically, global warming might alter the incidence of floods, droughts and water-related disease and ill-health; the threat of such changes might require some form of water management response.

There have been very few quantitative studies of the potential effect of climate change in Britain. The assessment presented here is based on qualitative evaluations derived from the quantitative *hydrological* studies outlined in previous chapters: some of these assessments will be vague. A context to the changes in Britain is given by reviews of studies undertaken in other countries.

THE MANAGEMENT OF WATER RESOURCES IN BRITAIN

Three groups are involved in water management in Britain: *suppliers, users* and *regulators*. The suppliers in England and Wales are the publicly-quoted and privately-owned water utilities. A few smaller local utilities are responsible just for the supply of water, but the ten larger regional utilities both supply water and treat effluent before discharging it into watercourses. These large regional utilities were formed from public water authorities in 1989 and the effects of this privatisation are still working through the water management and supply system. In Scotland, water is supplied by three public water authorities which took over from the Regional Councils in April 1996.

The principal regulator in England and Wales is the Environment Agency, which took over the functions of the National Rivers Authority (NRA) in April 1996. The NRA was formed in 1989 when water supply and treatment were separated from the old public water authorities. The NRA (and subsequently the Environment Agency) terms itself the "guardian of the water environment" and regulates other activities in the water sector as well as having an executive function. Its activities are divided into seven areas, although the first three are individually considerably larger than the last four together:

- Water resources (licensing abstractions and effluent return and ensuring availability of adequate resources);
- Flood defence;
- Water quality (standard setting and enforcement);
- Fisheries;
- Recreation;
- Conservation, and
- Navigation.

Figure 8.1 shows the operational regions of the Environment Agency.

The Department of the Environment (DoE) oversees the Environment Agency and, through the Office of Water Services (OFWAT), regulates the private water supply utilities. The Drinking Water Inspectorate ensures, on behalf of the Secretary of State for the Environment, that the water supply companies

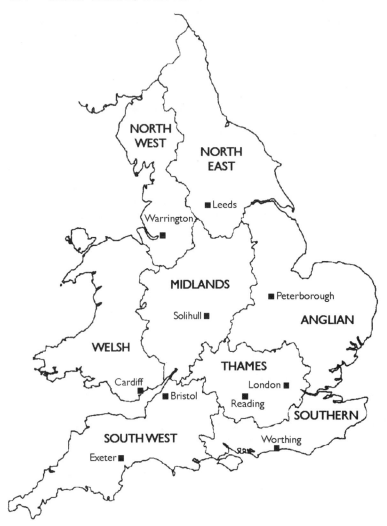

FIGURE 8.1 Operational regions of the Environment Agency

in England and Wales fulfil their obligations with regard to the quality of drinking water. The Ministry of Agriculture, Fisheries and Food (MAFF) provides grant aid to the Environment Agency for flood alleviation works. In Scotland the Scottish Environment Protection Agency (SEPA) has powers to protect against flooding and pollution.

The users of water are most obviously the domestic, municipal and industrial customers of the water utilities, but there are several other important users in Britain, many of whom abstract their own water. Some industries directly abstract water for use either in production or for cooling, and power generators abstract water for cooling. There are several major hydropower reservoirs in Scotland, operated by a recently-privatised utility, and a few reservoirs in Wales are also used for power generation. In England hydropower is largely confined to a small number of small-scale run-of-river installations. Hydropower plants generate less than 2% of the UK's electricity, with 97% of this coming from large-scale schemes (DTI, 1995). Irrigators and fish-farmers abstract water from rivers and groundwater. Navigators use water mainly via the canal system managed by the British Waterways Board (BWB). Finally, and increasingly importantly, water is a recreational resource (not least for fishing) and the demands of the aquatic ecosystem are being given increasing recognition. The role of the regulators, particularly the Environment Agency, is to balance the competing demands of all these users.

Climate change is likely to impact upon suppliers, regulators and users of water in different ways through changes in water resources. The water resources available for use in a region are determined by the *quantity* of water, the *quality* of water and the *demands* for water (Arnell, 1995a). Potential changes in the quantity and quality of water have been examined in detail in this book: before exploring possible impacts of global warming on the water sector and on environmental risk, it is necessary to examine potential changes in the demand for water.

POTENTIAL CHANGES IN DEMAND FOR WATER

A change in climate has the potential to affect the demand for water, particularly offstream demands. Most obviously, agricultural demands are sensitive to climate change, but domestic, industrial and general municipal demands may also alter. In most cases, climate change would be superimposed on other factors affecting per capita and gross demand for water, including population growth, economic development and technological change. Some of these factors, particularly technological change, are reducing the demand, but others are increasing it.

Figure 8.2 shows the proportion of total abstractions in England and Wales in 1992 by sector (Water Services, 1994). By far the largest quantities are abstracted for public water supply and electricity generation. Most of the water used in electricity is taken from Welsh rivers and much is returned to the river system downstream of the abstraction point, often at a higher temperature. Spray irrigation accounts for less than 1% of total abstractions, and this is largely concentrated in the Anglian and Midlands regions. Abstractions for other agricultural purposes make less than 0.5% of the total abstracted, but fish farms abstract almost 10% of the total: most of this water is returned rapidly to the river system. Industry abstracts approximately 6.5% of the total.

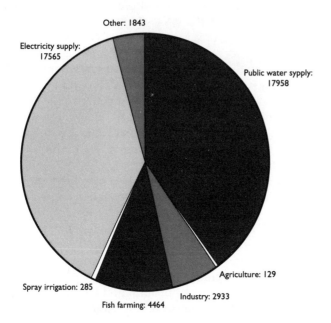

FIGURE 8.2 Proportion of abstractions in England and Wales by sector (data from Water Services Association, 1994)

Domestic, commercial and municipal demand

Demand in Britain

Domestic, commercial and municipal demand in Britain is largely supplied by the water utilities — 99% of the population is connected to public supplies (Water Services Association, 1994). The average domestic consumption is 147 litres per day per person, broken down as shown in Table 8.1. Total domestic, commercial and municipal demand in England and Wales increased by around 11% between 1973 and 1994 (Water Services Association, 1994), as a result of an increase in the number of households, an increase in appliances using water and (particularly) an increase in the use of water in the garden — a garden sprinkler can use 170 litres in ten minutes. Demand *per capita* is predicted to increase in southern England by 21% between 1990 and 2021 without climate change (Herrington, 1996), and the NRA forecast a growth of 12% in total domestic, commercial and municipal demand in England and Wales, under a medium growth scenario, by 2021 (NRA, 1994). Most of this growth is expected in the Environment Agency operational regions in southern and eastern England.

Impacts of climate change: the international context

There have been surprisingly few studies of the effect of climate change on demand for water. Cohen (1987b) developed regression relationships between temperature, soil moisture deficit and total municipal withdrawals in the Great Lakes region of North America, and found an increase of just over 5% with a temperature increase of between 3.1 and 4.8 degrees Celsius.

Impacts of climate change on domestic demand in Britain

Domestic demand for water in Britain is sensitive to climate. During the warm, dry summer of 1995 peak demands were considerably higher than in previous summers, due largely to the watering of gardens. Herrington (1996) estimated the changes in the components of domestic demand and concluded that a temperature rise of just over 1°C by 2021 could add another 4% onto the 21% increase in per capita demand which is already likely to occur in southern England by 2021 (Table 8.1). This additional increase would largely be due to even greater use of water in the garden. Peak demands would be increased by a greater proportion. Table 8.2 shows the estimated ratios of the highest 7-day demand to the average 7-day demand (termed the peak 7-day ratio) in non-metropolitan south-east England in 1991 and 2021, for both domestic demand and for total public water supply. The changes were estimated by

TABLE 8.1 Domestic demand components for non-metropolitan south-east England: 1991 and 2021, with and without climate change (Herrington, 1996): litres per head per day.

Component	1991	2021 no climate change	2021 +1.1°C warming
WC use	35.5	33.6	33.6
Showering	5.3	24.0	26.8
Other personal washing	41.2	37.6	37.6
Clothes washing	21.7	22.0	22.0
Dish washing	11.8	11.0	11.0
Waste disposal unit	0.4	1.5	1.5
Car washing	0.9	1.5	1.5
Lawn sprinkling	2.5	8.7	11.7
Other garden use	3.8	7.2	8.6
Miscellaneous use	23.9	31.3	31.3
TOTAL DOMESTIC USE	147.0	178.4	185.6

TABLE 8.2 Peak 7-day ratios in non-metropolitan south-east England: 1991 and 2021 (Herrington, 1996)

Peak 7-day ratio	1991	2021 no climate change	2021 +1.1°C warming
Domestic peak	1.35 – 1.45	1.6 – 1.8	1.7 – 1.94
Public water supply system peak	1.16 – 1.27	1.27 – 1.45	1.315 – 1.52

assuming that current peak 7-day ratios for in-house and ex-house use (1.1 and 6.2 - 8.25, respectively) could be applied in the future to the altered in-house and ex-house totals. The predicted increase in peak demands puts great potential strain on the design and operation of water distribution networks.

Domestic demands constitute the largest single component of public water supply (44% of total public water supply, including leakage, in south-east England: Herrington, 1996), but other demands supplied by the water utilities through the public supply system include commercial demands, golf courses, parks and agriculture (except irrigation, which is supplied separately and is generally abstracted directly by the farmer). Demand from golf courses and parks would increase as potential evaporation rises, but the totals involved, although locally significant, are small relative to the other components of public supply. Commercial use is generally much less sensitive to temperature than domestic use, but does include the water used in air conditioning systems. Air conditioning is expected to be used more frequently following global warming, so demand for water from air conditioning systems may increase. However, for economic and public health reasons, future air conditioning systems would largely be air cooling rather than chilled-water systems, so the effects on water demand may be trivial.

Changes in the total demands for public supplies are dominated by changes in domestic demand. As indicated above, Herrington (1996) estimated a 5% additional increase in *per capita* demand by 2021 due to global warming. He translated this into an additional 5.5% increase in *aggregate* domestic demand in the south and east by 2021, due to the increasing population, and an additional 4% increase in total demand for public water supply: this is equal to 230 Ml d^{-1}, or 37% of the water supplied in 1993/94 by Southern Water. Total public water supply peak 7-day ratios increase by a smaller proportion than domestic peak ratios (Table 8.2) because the other sources of demand which make up public water supply are less variable over time.

Finally, when considering public water supply it is important to recognise that almost a quarter of water in Britain is lost through leakage from the distribution network (DoE, 1992). Some of this leakage is through pipes which freeze and burst: this may become less significant as temperatures rise and freezing episodes become less frequent — the frequency of *air frosts* may halve by the 2050s (CCIRG, 1996).

Demand for industrial and cooling water

Industrial use of water is varied, ranging from water used in food and raw material processing and manufacturing, to water used for cooling industrial processes. Water use by industry is increasingly efficient, particularly as the price of water increases, and much industrial demand will be unaffected by climate change. Those industries that are most dependent on water, such as food processing, will be more affected by changes in the availability and price of water.

The vast bulk of the water abstracted by industry in Britain (85%) is used for cooling, particularly by power generators. Changes in the demands for power will affect demands for cooling water and a rise in water temperature will also affect demand because cooling efficiency declines as temperature rises. However, technological changes in the electricity generating industry, such as the introduction of combined cycle gas turbine (CCGT) designs which have much lower cooling water requirements than traditional steam turbine designs, will lead to a reduction in the demand for cooling water over the next few decades.

Agricultural demand

Both rain-fed and irrigated crops would require more water if temperature and evaporation were to increase, although this might be offset by the effects of increased rainfall. An increased consumption of water by rain-fed crops would appear to reduce the amount of rainfall reaching the stream network, but frequently during the peak demand season evaporation is limited by moisture availability. An increase in demand by rain-fed crops might not therefore have a major effect on runoff.

An increase in demands from irrigated crops could have greater significance because crop yield and quality can be dependent on demands for water being met. Allen *et al.* (1991) investigated potential changes in irrigation demand in the Great Plains region of the USA, using the Penman evaporation formula. They simulated increased demands from alfalfa, due largely to increases in the length of the growing season and increased requirements during summer. Irrigation demands from winter wheat and corn were increased by a smaller amount and, in some cases, demand fell due to reductions in the length of crop life cycles and the time taken for the crop to ripen; changes in plant physiology also tended to compensate for increased temperature. Peterson and Keller (1990) calculated the Net Irrigation Requirement (NIR) across the United States, under a range of scenarios for change in temperature and precipitation. The NIR is the amount of water needed to maximise crop production at a particular location, and equals crop evapotranspiration less usable rainfall. Crop requirements for each study county were based on standard cropping patterns, with growing seasons and crop scheduling patterns adjusted in line with altered temperatures. Under all the scenarios considered, NIR increased, with the amount of increase depending on the change in precipitation everywhere but arid regions: here precipitation changes are largely irrelevant, because precipitation is always small relative to potential evaporation. Over 39 sample counties, the NIR increased by 15% with a 3°C increase in temperature and no change in precipitation; with a 10% increase in precipitation added, the increase in NIR would be 7%, whilst if precipitation were to fall by 10%, the NIR would increase by 26%. Cohen (1991) also simulated an increase in irrigation demands, this time in the Saskatchewan river basin, Canada, using empirical relationships between irrigation demands and soil moisture deficit. With a 2°C increase in temperature and a 10% decline in rainfall, total irrigation demands in the Saskatchewan basin would increase by between 39% and 133%, depending on the rate of growth of irrigated areas over the next few decades (illustrating, incidentally, how the effects of climate change may be small compared with the effects of other, non-climatic, change).

Future irrigation demands will depend on climate and factors such as demand for irrigated produce, agricultural prices, the price of water and the efficiency with which water is used in irrigation. Although there has been rapid expansion in irrigation over the last few decades as demand for food has risen, there is likely to be only limited future expansion in irrigated area at the global scale as many of the most economic locations have already been exploited (Postel, 1993). Expansion is possible where irrigation is geared towards high-value produce, but in general it is likely that improvements in the efficiency of irrigation water use will lead to the greatest changes in irrigation demands.

The demand for spray irrigation in Britain has increased over the last two decades, and is predicted to increase further in the future. The Environment Agency forecasts an increase in demand for spray irrigation of 75% between 1990 and 2021 (Weatherhead *et al.*, 1994). Herrington (1996) developed a regression model between de-trended irrigation abstractions and summer temperature from 1969 to 1990 in the Anglian region. Variations in temperature explained approximately a quarter of the variations in de-trended irrigation demands and an increase in temperature of just over 1°C by 2021 would lead to an increase in abstractions of 28% or 155 Ml d^{-1} in southern England. When this climatically-induced increase is combined with estimates of future irrigation demand, the total demand by 2021 is predicted to increase by approximately 125% over 1990 levels (Herrington, 1996). This increase will be focused on a few areas, and concentrated in summer.

Spray irrigation represents approximately three-quarters of the water used by agriculture in southern England. The remaining quarter is largely supplied through the public supply network and is mostly used for stock watering. A regression analysis between annual consumption by a few large agricultural consumers (with trend removed) and temperature indicated that a rise in temperature of just over 1°C by 2021 would give an increase in demand of approximately 12%, or 18 Ml d^{-1} in southern England (Herrington, 1996): a further rise of 18% over the same period is likely for reasons other than climate change.

Aggregate demand

Table 8.3 gives Herrington's estimates of present and future aggregate demand for water in south and east England, excluding leakage. The increase in public water supply is dominated by the increase in domestic demand, with increases in spray irrigation by far the next largest. The table emphasises the uncertainties in the estimated future demands for water (represented by the ranges given in parentheses), and places the possible effects of climate change alongside the effects on demand of other changes. These changes, such as population

TABLE **8.3** Present and future demands for water in the south and east, 1991 and 2021: Ml d^{-1}; range in parentheses (Herrington, 1996)

	1991 Best estimate (Ml d^{-1})	2021 Without climate change (Ml d^{-1})	2021 With climate change (Ml d^{-1})	Increase due to climate change (Ml d^{-1})	Increase due to climate change (%)
Public water supply					
Domestic	3740	5069 (4562-5575)	5273 (4746-5800)	204	5.5
Air conditioning	16 (7-26)	16 (7-26)	22 (9-39)	6	37
Golf courses	3.3	4.8 (4.3-5.3)	5 (4.5-5.5)	0.2	6
Other parks	24 (16-33)	30 (19-43)	32 (20-45)	1	5
Agriculture (except irrigation)	125 (88-162)	154 (98-218)	172 (110-244)	18	15
Other commercial [1]	1770 (1715-1824)	1903 (1715-2098)	1903 (1715-2098)	0	0
Unmeasured and miscellaneous [2]	432	464 (432-497)	464 (432-497)	0	0
Total public water supply (excluding leakage)	6111	7641 (6838-8462)	7871 (7037-8728)	230	3.8
Spray irrigation					
Total	333	562 (418-706)	717 (533-900)	155	46

[1] excluding air conditioning use   [2] excluding leakage

growth, improvements in the water-use efficiency of appliances and irrigation and the effects of altered pricing structures, are also very uncertain so forecasts for increases in the demand for water are notoriously unreliable.

The effects of changes in instream water demands (particularly recreation and ecosystem demands) are considered in later sections.

IMPACTS ON WATER RESOURCES

The previous sections have summarised how the amount of water, the quality of water and the demand for water might change with global warming; this section examines implications of these changes for water uses and users.

Water supply

Sensitivities of supply systems to changes in climate

There are three important characteristics of a water supply system. The first is its total storage capacity, relative to inputs. A direct river abstraction has zero storage whilst reservoir and groundwater supply systems have a storage capacity. The second important system characteristic is its yield, again relative to the size of inputs or storage volume. The third characteristic is the probability of "failure", defined in the most general sense as the probability of not meeting supply targets.

For a given size of storage, a change in inflows will lead either to a change in the risk of failure, or to a change in the yield that can be extracted with a given reliability: the degree of change in risk or yield depends on the volume of storage (relative to inflows). In general, the smaller the volume of storage, the greater the relative change in yield or risk for a given change in inputs.

Several studies have examined the sensitivity of hypothetical supply systems to changes in climatic inputs. They fall into two groups: the first take an analytical approach, and estimate changes in system reliability given changes to the probabilistic structure of inflows. Figure 8.3, for example, shows the reliable yield (the yield exceeded 90% of the time) of a reservoir under different changes to the mean and standard deviation of inflows, and changes in the amount of evaporation (Rogers and Fiering, 1990). The base case is at G, and line AB describes the effects of a change in just the mean. Line CGD shows the effect of a change in just the standard deviation of inflows (both an increase and a decrease would lower yields), and EF shows the combined effect of changing the mean and standard deviation. Line GH shows the effect on yield of just increasing evaporation.

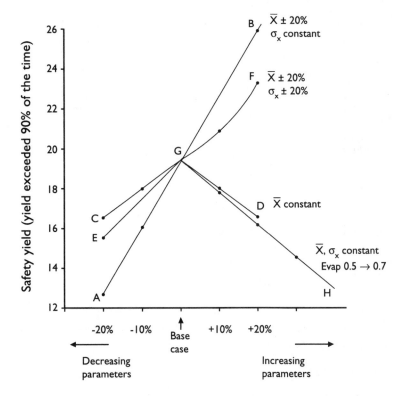

FIGURE 8.3 Sensitivity of yield to climate change: analytical approach (Rogers and Fiering, 1990)

The second group of studies has taken a simulation approach, putting simulated flow sequences through a simple reservoir model. Nemec and Schaake (1982), for example, found that a given percentage change in inflows would have a greater relative effect on yield or the storage required to maintain yields. In their humid example, a reduction in rainfall of 25% would necessitate an increase in storage of 400% in order to maintain existing yield and risk of failure. Cole *et al.* (1990) simulated hypothetical reservoir reliability in south-east and north-west England, and plotted storage-yield curves under current and changed climates (Figure 8.4). These curves show the yield that can be obtained with a given reliability, from a reservoir with a given storage volume. For example, in order to keep the same reliability of supply, yield would have to be reduced from A to B, or storage increased from A to C, under the change scenario examined. An 8% reduction in annual runoff, as simulated under the scenario, would lead to reductions in yield of between 8 and 15%, depending on reliability, or increases in storage requirements of between 10 and 21%.

A reservoir reliability analysis, similar to that of Cole *et al.* (1990), was carried out for hypothetical reservoirs in two of the case study catchments: the Harper's Brook in Northamptonshire and the Teme in the West Midlands*. A simple behaviour

---

* This does not imply that reservoirs are proposed for these catchments

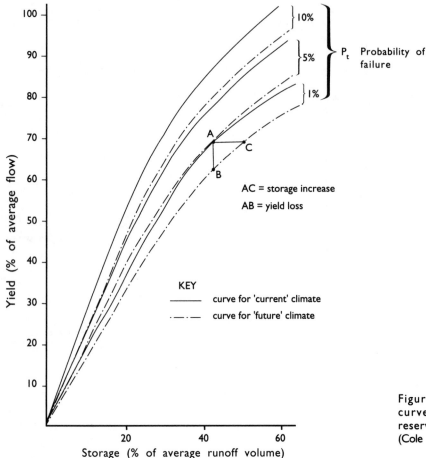

Figure 8.4 Storage-yield curves for a hypothetical reservoir in south-east England (Cole *et al.*, 1990)

analysis (McMahon and Mein, 1986) was performed with different storages and yields, both expressed as a proportion of baseline runoff. Simple behaviour analysis assumes that yield is maintained until the reservoir empties and reliability is measured in terms of the risk of the reservoir being unable to supply the yield in any one month (a reliability of 5%, for example, means that there is a 5% chance that the reservoir will not be able to supply the yield in any given month). It is unrealistic in the sense that it ignores actions which would be taken by reservoir managers in the face of shortage, and thus overstates the risk of "failure", but gives an indication of potential changes in reservoir reliability and the effect of storage volume on sensitivity to change.

Figure 8.5 shows the reliability of supply for a reservoir in the Harper's Brook catchment with yields of 50% and 100% of baseline average flow, for different storages and under baseline and altered climates. With the lower yield (50% of baseline average flow) there is no difference in reliability for the largest storages: if the storage volume is greater than 35% of the baseline annual runoff volume, then the reservoir could maintain the yield following climate change with no reduction in reliability. However, with the larger yield, there is a larger proportional change in reliability and even with very large storage volumes there would still be a reduction in reliability after climate change. Figure 8.6 shows the same for the Teme catchment. The broad pattern is similar, although the change in

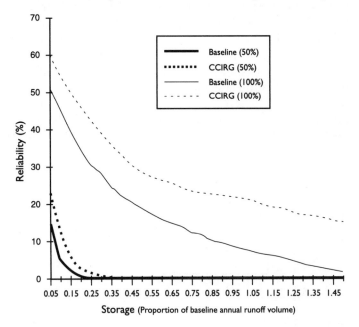

FIGURE 8.5 Reliability of supply for a hypothetical reservoir in the Harper's Brook catchment: yeilds of 50% and 100% of baseline average flow

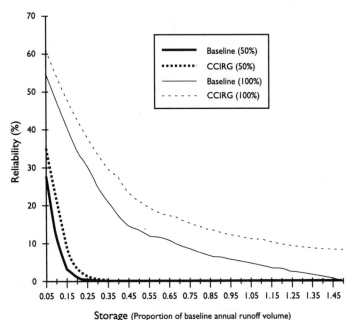

FIGURE 8.6 Reliability of supply for a hypothetical reservoir in the Teme catchment: yeilds of 50% and 100% of baseline average flow

reliability between the baseline and altered climates is smaller. Figure 8.7 shows the storage necessary to supply a given yield with a reliability of 5%, under baseline and changed climates, for both catchments. In general, the required storage is higher under the 1996 CCIRG scenario, with the greater the yield, the greater the proportional increase needed.

These types of studies give general insights into the sensitivity of water supply systems to climate change, and show that changes in yield or storage requirements may be considerably greater (in relative terms) than changes in climate inputs, and that the effect of a change in climate depends significantly on current storage and yield relative to inflows. The problem with this type of study is that real-world supply systems are usually much more complicated and impacts of climate change are not necessarily so clear. Real-world systems are often interconnected, exploiting a variety of sources, and are usually managed so that complete failure during drought conditions is averted.

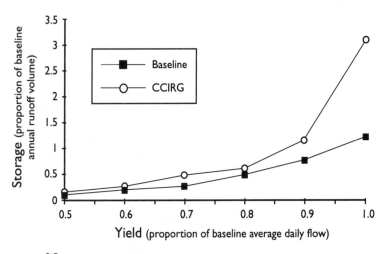

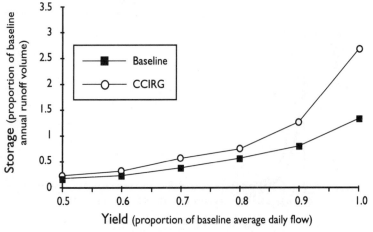

FIGURE 8.7 Storage-yield relationship for a reliability of 5%: Harpers Brook (top) and Teme catchments

Impacts on specific supply systems: some international examples

Wolock *et al.* (1993) describe an investigation into the Delaware River system in the eastern United States. The Delaware River basin has an area of 30 000 km², and supplies water to 7 million people living within the basin (including Philadelphia) and to a further 13 million people outside, largely in New York City. New York City is supplied through a network of reservoirs in the upper part of the catchment and other supplies are taken from the Delaware River, close to the saline limit. Water management in the basin is a compromise between extracting water to export to New York City and ensuring enough water is left in the river to prevent salt water infiltrating the intake points in the lower reaches. Target discharges are set for specific points along the river, constraining the operation of the upstream reservoirs. Wolock *et al.* (1993) used a water balance model to simulate flows at various points in the Delaware basin and fed the simulated flows through a model of the reservoir system to determine both the proportion of time that the contents of the New York reservoirs fell below a critical level and the position of the salt front in the lower reaches relative to the Philadelphia intakes. Transient arbitrary climate change scenarios were applied, assuming gradual change over a 100 year period. The New York reservoir system was found to be considerably more sensitive to changes in climatic inputs than Philadelphia supplies, and Figure 8.8 shows the percentage change in "crisis" conditions for both systems with changes in temperature and precipitation: a reduction in precipitation would have very significant effects on the New York supply system. Wolock *et al.* (1993) also simulated the effects of reducing the target discharges. Impacts on the New York system were reduced — because the effective storage of the reservoirs had

FIGURE **8.8** Sensitivity of water resources in the Delaware basin to changes in precipitation and temperature: 2°C warming (Wolock *et al.*, 1993)

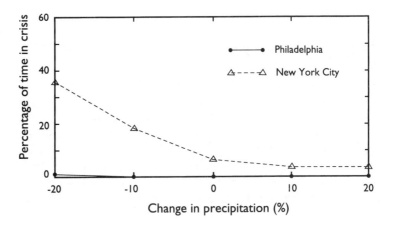

increased — and there was very little effect on the position of the salt front: this was probably because lower releases from the reservoirs meant that more water was available to maintain flows during critical dry periods. Wolock *et al.* (1993) concluded that decreasing target flows reduced the sensitivity of water resources in the Delaware basin to climate change.

Kirshen and Fennessey (1995) investigated possible impacts of climate change on the system supplying water to Boston, Massachusetts. The main sources of supply are two reservoirs, fed from three river basins to the west of Boston. The safe yield of the system (with a reliability of 98.5%) is 1054 Ml d⁻¹, and demand was only reduced to below this figure in the late 1980s by aggressive demand management. A water balance model was used to simulate flows in the source catchments, and a water resources model was used to simulate the transfer of water between catchments and reservoirs. Four different GCMs were used to define climate change scenarios, all representing equilibrium $2 \times CO_2$ climates. Two of the scenarios assumed an increase in precipitation, whilst two assumed a decrease: all assumed an increase in potential evaporation. Table 8.4 shows the change in evaporation, precipitation, river flows (summed across the three source catchments) and safe yield (reliability of 98.5%), under the four GCM scenarios. The percentage change in yield is greater than the percentage change in river flows. Under the GISS scenario, the reliability of the current safe yield (1054 Ml d⁻¹) falls from 98.5% to 83.4%, and drought management practices would only increase the reliability to 86%.

	Precipitation	Potential evaporation	River flow	Safe yield
GISS scenario	-1.6	+20	-16	-23
GFDL scenario	-7.6	+57	-33	-43
OSU scenario	+13	+23	+6	+7
UKMO scenario	+32	+10	+30	+38

TABLE 8.4 Percentage change in precipitation, potential evaporation, river flow and system safe yield, under four GCM scenarios: Boston water supply (Kirshen and Fennessey, 1995)

The effect of climate change on water supply and distribution systems in Britain

Public water supply in England and Wales is provided by ten regional water service companies and 21 (in 1995) local water supply companies, who together provide virtually all domestic, commercial and municipal supplies, agricultural supplies other than irrigation, and supplies to some industrial users. Water to these users in Scotland has been provided, since April 1996, by three public water authorities. This section focuses on such suppliers to the public, and their networks of supply and distribution systems. Most industrial abstractors and irrigators take water directly from rivers or boreholes for their own use.

The public water supply system in Britain is essentially divided into three parts: the provision of water resources, the distribution of water to consumers and the treatment of water once it has been used by consumers. Water supply companies in England and Wales operate under agreed standards of service contracts with the regulator OFWAT. One set of targets for each company relates to the frequency of restrictions on users. For example, OFWAT targets are for hosepipe bans no more than once in ten years, major publicity campaigns for voluntary reductions once in twenty years, and rota cuts or standpipes at most once in a hundred years (DoE, 1992). In order to meet these targets in future, water companies need to consider the potential effects of climate change on resources, distribution and treatment.

Water resources
In England and Wales three water sources are used approximately equally:
- Direct abstraction from rivers;
- Abstraction from water stored in reservoirs, and
- Abstraction from groundwater.
There are few abstractions from groundwater in Scotland.

The flows from some of the rivers used for abstraction are supported by controlled releases from upstream reservoirs or by transfers from other basins. A system for the supply of water in a region may consist of a number of independent supply sources or, at the other extreme, a network of connected sources which can be mixed and merged according to cost and availability. The Lancashire Conjunctive Use System, supplying large parts of north-west England, is an example of an integrated system including groundwater, direct river abstractions and reservoirs. The size, or yield, of the sources within the supply system is determined by the gross volume of water demanded, averaged over months or seasons. A change in climate might affect the yields of the various sources within a system, with the effects on system yield depending on how the system is designed and operated, and might also affect the gross demands placed on the system.

Direct river abstractions would be adversely affected during summer in southern Britain, where summer river flows reduce under the 1996 CCIRG scenario. This would affect some public water supplies, but would also impact upon those users — particularly irrigators (see below) — who abstract their own water. Higher water temperatures and poorer water quality in some rivers might lead the regulators to impose higher minimum flow constraints on abstraction licences, further reducing the amounts which could be withdrawn. In northern and upland

Britain, where summer flows under the 1996 CCIRG scenario increase, direct river abstractions are likely to be less affected.

Reservoir yield would be affected not only by the size of the climate change but also by the characteristics of the reservoir system. As indicated above, small reservoirs with little storage are likely to be most affected by change and a supply system relying on many small reservoirs that are frequently filled and emptied would be more sensitive than a system relying on one large reservoir (unless the timings of fill and draw-down in the small reservoirs were so different that each could be used at different times). Small systems, relying on frequent replenishment, would be particularly seriously affected by a reduction in spring and summer inflows — as occurred in Yorkshire during the drought of 1995.

Hewett *et al.* (1993) simulated changes to the yield of the Weir Wood reservoir in the Medway catchment in Sussex, under several change scenarios similar to, but wetter than, the 1996 CCIRG scenario. At present, the reservoir with a storage of 5500 Ml (approximately 50% of the annual runoff volume) can supply a yield of 13.9 Ml d^{-1} (approximately 50% of the average daily flow) with a reliability of 98%. Under two scenarios with increased winter flows and reduced summer flows, the yield of the reservoir would *increase* by between 4 and 12%, because the reservoir would stay full for longer in autumn and spring. With reduced spring inflows, reservoir yield would decrease.

The use and effectiveness of inter-basin transfers, such as that from the Ely Ouse to the Essex Stour in eastern England, will depend on changes in both the receiving and the supplying water course. In most cases, transfers only operate to sustain flows during low flow periods on the receiving river and are conditional on the amounts of water available in the supplying river. An increase in the frequency of low flow spells, as implied across much of Britain under the 1996 CCIRG scenario, would suggest increased demands for inter-basin transfers but the ability of the source rivers to provide water may also be affected. An increased regional variability in resources might increase the potential for inter-basin transfer. As with reservoir systems, the precise effect of climate change will depend on the operating rules and, more particularly, the volumes transferred relative to the total flows in the source river.

Reduced groundwater recharge and lower groundwater levels would affect yields from boreholes, depending not only on the magnitude of change but also on any restrictions imposed by the licensing authority on borehole abstraction rates. For example, abstractions may be constrained by the need to maintain river flows within the catchment, and this might exaggerate the effects of changing recharge on borehole yields.

There are many possible options for adaptation to climate change, and the water industry in Britain is continually planning to meet altered circumstances (usually, higher demand). Water resources management options essentially fall into two categories, and both can in principle be used in response to climate change. Supply-side options aim to provide more resources to consumers, either through new resource developments or through alterations to existing schemes. Demand-side options aim to close the gap between supply and demand by reducing demand for water. They include efforts to reduce consumption by users, as well as reductions in the amount of water lost through leakage. The NRA conducted a review in 1994 of possible ways of meeting the increase in demand expected in the absence of climate change (NRA, 1994). The review listed a number of possible local resource development options — generally involving small schemes or modifications to existing schemes — and explored several strategic options (Figure 8.9) to meet different demand scenarios: the different scenarios made assumptions about consumption of water, the effect of water metering on demand, and future

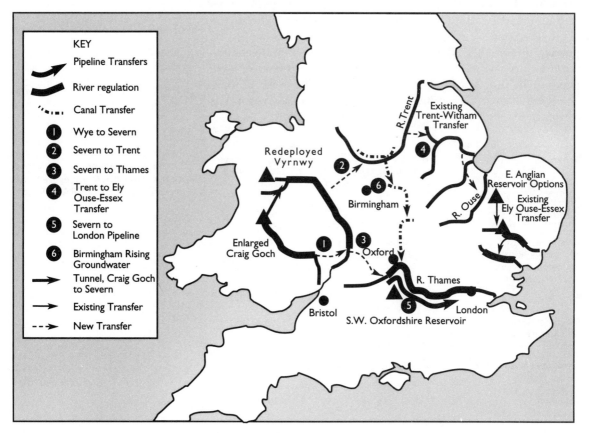

FIGURE 8.9 Strategic options for meeting increased demand (NRA, 1994)

reductions in leakage. The large range in total demands between scenarios (a difference of 2000 Ml d^{-1}, or approximately 15% of current demand in England and Wales) indicates the potential for demand management to curb future water shortages.

Most of the strategic supply options involve a transfer of resources between basins, although some involve reservoir development. A number of options were considered but excluded, largely on environmental grounds. These include tidal barrages, artificial recharge of groundwater, effluent reuse, desalination and bulk shipments by tanker. There are many problems with the implementation of new resource developments: these are summarised in the NRA review (NRA, 1994). The most obvious one is cost: a reservoir in south west Oxfordshire would cost £400M, for example, whilst transfers from the Trent to the Anglian regions could cost £108M. Also important, however, are environmental impacts and political considerations. Environmental impacts include not only the loss of land to reservoirs, but also the instream effects of adding water with different physical characteristics into a river system. Nevertheless, the Environment Agency has identified potential responses to shortfalls in demand, and these responses can be implemented to adapt to climate change. Table 8.5 shows the estimated time by which various strategic options would be needed under a high demand growth scenario: climate change would "simply" alter the time by which new resources are needed. In practice the uncertainty associated with climate change makes any assessment of the timing of new resources difficult — a point returned to later in the chapter.

Scheme	To meet new demands in...	Approximate time needed
Unsupported River Severn to River Thames transfer	Thames region	1996
Partial development of Vyrnwy Reservoir to support Severn-Thames transfer	West Midlands, Wessex and Thames	2001-2006
East Anglian Reservoir	Anglian region	2001-2006
Enlarged Craig Coch, to support Severn-Thames transfer	West Midlands, Wessex and Thames	2006-2011
River Severn to River Trent transfer	East Midlands	2011-2016
South-west Oxfordshire reservoir	Thames region	2016-2021
Birmingham rising groundwater	East Midlands	2006-2011

TABLE 8.5 Timing of new resource developments, required under the NRA high demand growth scenario (NRA, 1994)

Distribution to consumers

The second component of the supply system is concerned with water treatment and distribution from source to consumer. This comprises a network of water treatment works, service reservoirs and water mains pipes. The size of the distribution network — capacity of treatment works, service reservoirs, dimensions of mains pipes — is determined by the peak demands placed on the system, rather than by long-term volumes demanded. In the south and east these peaks tend to arise during summer, whilst in the cooler north and west they occur following pipe bursts in winter and spring. A change in climate would affect the performance of the distribution network largely through effects on peak demands. As indicated above, these effects may be substantial. The high demands during the hot summer of 1995 showed how sensitive distribution networks can be to changes in peak demands.

The impacts of an inadequate distribution network would be manifested in restrictions in water use by consumers. In less extreme circumstances, consumers would experience a reduction in pressure during peak periods. Under more extreme conditions, the water supply companies encourage restraint in water use and enforce restrictions on certain types of use. For example hosepipe bans tend to have the specific objective of reducing peak demands, but they are unpopular and OFWAT only permits them once every ten years.

To prevent distribution problems caused by increased peak demands, water supply companies will probably need either to invest in an upgraded distribution infrastructure or to develop demand management techniques to curb peak demands. Investment in larger capacity distribution networks may be very expensive, particularly where this involves renovating buried water mains. Demand management techniques for curbing peak demands are currently being explored by the water industry, but could include higher tariffs for water used either at certain times or above certain thresholds: this would require metering.

Treatment of effluent

The third component of the supply system is the treatment of water once it has been used by the consumer. In England and Wales 95% of water supplied to consumers (excluding leakage) is returned as effluent to be treated by the ten regional water service companies, then discharged to water courses or directly into the sea. The capacity and standards of sewage treatment works are determined by the quality of discharges and receiving water as specified by the Environment Agency. A change in climate would affect the operation of sewage treatment works primarily by altering the characteristics of the receiving water. For example, if the initial dissolved oxygen concentration is reduced, the receiving water will be able to accept a smaller quantity of effluent discharged at the current quality. The volume of effluent which can be discharged would have to be reduced and the capacity of the treatment plant would decline. However, higher temperatures should make the biological processes in sewage treatment works operate more efficiently, and so treat effluent to required standards more rapidly.

Implications of climate change for water supply: some general conclusions

The effects of changes in climate on public water supply are complicated and depend significantly on the characteristics of the supply system and how it is operated. In the most general sense, it is likely that changes in the volume of available water and the gross demand for water will affect the resource base, changes in peak demands will have the greatest effect on the distribution network, and changes in stream water quality will have implications for effluent treatment and the cost of returning treated water to the water course. In Britain, under the 1996 CCIRG and other feasible scenarios, it is likely that the effects of global warming on supply and demand would put water resources under increased pressure. There would be increased regional variation in the resource base, with the greatest extra pressures in the south and east — the regions with the greatest strain on water resources in dry years already. These judgements, however, are not based on quantitative studies, and there is a clear need for studies into the possible financial and practical impacts of climate change on water resource systems. Are the possible changes within the design capabilities of existing and planned systems, for example, or would they pose stresses which could only be managed through expensive capital works? More importantly, how do the possible effects of climate change compare with the effects of other changes, such as growth in demand, which are already being planned for?

There have been very few comprehensive studies worldwide into the effects of climate change on real water resources systems, considering not only changes in water availability, but also changes in demand for water, the effects of exogenous (non-climatic) changes and the effects of progressive adaptation to change.

Agriculture and irrigation

Agriculture uses water for many purposes, including stock watering, equipment cleaning and irrigation. Some of this water may come from public supplies, some may be abstracted directly by the farmer and some may come from large agricultural water supply projects covering many farms. The proportions of these different sources vary between regions. In the American west, for example, major agricultural water supply schemes provide water to large numbers of farmers. These large supply systems use networks of reservoirs and distribution systems and are in many ways analogous to the public supply systems discussed in the previous section. In most of western Europe virtually all water used on farms is taken either from the public supply network or is abstracted by the farmer from rivers or boreholes. In this case, climate change will affect farmers through changes in public supply and, more directly, changes in the reliability of their own supplies.

Several studies have looked at demand for irrigation water but there has been very little work on potential changes in the ability of an irrigation system to meet demands. Arnell and Piper (1996) simulated behaviour of a hypothetical irrigation system in Lesotho, accounting for changes in both demand and supply. They found substantial changes in the reliability of the irrigation system. An increase of temperature of 2°C, for example, would increase demand for irrigation water by over 20%, whilst a reduction of rainfall of 10% would mean that the irrigation reservoir would be unable to supply irrigation demands two out of every three years, rather than one in five as under the present climate. By simulating a reduction in irrigated area as water shortages developed (a form of adaptation to changing conditions) they found that for a temperature increase of 2°C, the percentage of years in which the area irrigated was less than the target would rise from 14% to 50%, and the mean area irrigated would reduce.

Supplying water to agriculture in Britain

As already discussed, three-quarters of the demand for water from agriculture in Britain is for spray irrigation, the demand for which is very seasonal and highly concentrated in eastern England. Virtually all of this demand is met by farmers themselves, either abstracting directly from watercourses or aquifers, or taking water from on-farm ponds filled during winter. During warm, dry summers, irrigators may be prohibited from abstracting water from rivers. This occurred in parts of East Anglia during 1991 and 1992. Reduced summer flows in eastern and southern Britain would greatly reduce the ability of farmers to irrigate their crops. Over the next few years the way irrigation water is abstracted is likely to change as irrigation becomes more critical. More farmers will construct on-farm storages (already required by the Environment Agency before new abstraction licences are granted) and demand will shift from summer to winter. Increased winter flows, as simulated under the 1996 CCIRG scenario, would make such a strategy more favourable.

The remaining quarter of agricultural use in Britain is supplied through the public supply network so would be affected by changes in the reliability or price of water.

Power generation

Hydroelectric generation

By the beginning of the 1990s, hydroelectric facilities provided approximately 24% of world electricity generating capacity (Gleick, 1993). The distribution of hydroelectric installations around the world is very uneven. North America, for example, has developed approximately 60% of its large-scale hydropower potential, whilst Europe has developed 36% and Africa only 5% (Gleick, 1993).

There are various types of hydroelectric generation plants: all of them use falling or moving water to turn turbines (water mills, which use water to do mechanical work, represent a form of "hydropower"). There is a distinction between schemes which involve a dam to store water and provide a drop (or "head") for water to fall, and run-of-river schemes, which take advantage of the continuous flow of water past the turbine to generate electricity. The height of a hydroelectric dam determines the amount of electricity that can be generated, whilst the volume of storage controls the sustainability of power production. Hydroelectric dams can be very large, and include the largest dams and reservoirs in the world. Run-of-river schemes do not involve the construction of a dam, although many take advantage of weirs or locks in the river to provide a small amount of head. The generation potential of a run-of-river scheme is largely a function of the consistency of flow over time, and is related to the shape of the flow duration curve. In fact, power generation potential is the integration of the flow duration curve between the lowest flow at which turbines can turn (often around the flow exceeded 90% of the time, but depending on turbine efficiency) and the maximum flow which may be passed through the turbines (often the average flow). Schemes vary in size from large ones on the River Rhine, Germany, to micro-hydropower schemes on many small rivers throughout the world, generating small amounts of electricity for local consumers. A third type of hydroelectric scheme takes advantage of the head provided by a natural waterfall, such as at Niagara Falls, North America, where (unlike schemes with dams) there is no storage to sustain power generation during dry periods.

Climate change has two potential major effects on hydroelectric potential. The first arises from a change in inflows. In a run-of-river scheme, a change in the shape of the flow duration curve will affect the consistency of flows and hence the power potential. In a scheme involving reservoir storage, the effect of the change in flows depends on the capacity of the reservoir (in much the same way as reservoir capacity affects the impact of changes in inflows on water supply). The second potential effect of climate change is on evaporation from a reservoir surface, which may significantly reduce production potential.

There have been several studies into the effects of climate change on hydroelectricity generation, although these studies have concentrated on reservoir-based systems (e.g. Nash and Gleick, 1993). Mimikou *et al.* (1991) simulated the behaviour of a combined water supply and hydropower system in central Greece and calculated the risk of being unable to generate the design power under a range of arbitrary change scenarios. With a reduction in precipitation of 20%, the risk of one of the hydropower reservoirs failing to generate its design power rose from less than 1% to more than 20% (Figure 8.10). In Norway, Saelthun *et al.* (1990) simulated a substantial change in the distribution of power generation potential through the year, reflecting changes in reservoir contents (Figure 8.11). Lower snowfall, and hence more inflows to reservoirs during winter, would mean that production potential would be higher in winter — when most needed — and the surge of power potential in spring would be reduced. Haneman and McCann (1993) simulated changes in power production in northern California, finding a reduction in annual power production of 3.8%, but a decline during the peak demand period of up to 20%. They estimated that if the shortfall in power was made up

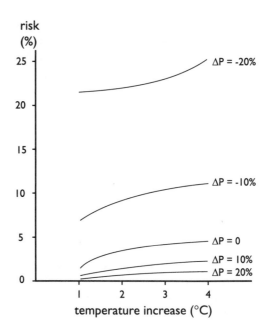

FIGURE 8.10 Risk of being unable to generate design power, under arbitrary change scenarios: Mesohora Reservoir, Greece (Mimikou *et al.*, 1991)

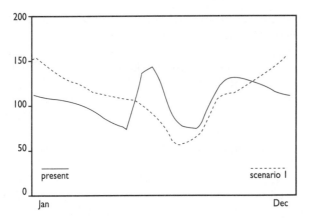

Figure 8.11 Distribution of power generation potential through the year, for a reservoir in Norway (Saelthun et al., 1990)

using natural gas, costs would increase by $145 million per year — assuming no other technical improvements or other responses to climate change.

In Britain, hydroelectric power represents less than 2% of total electricity generation, and is concentrated in a number of large reservoirs in Scotland. The effect of climate change on this production potential would depend on how the inflows to the reservoirs would alter, and under the 1996 CCIRG scenario it appears that flows in Scotland would be increased for much of the year. This implies that production potential would increase, although this depends on whether the reservoirs can store the additional water, and on whether an increased allowance would need to be made for flood protection. At present, hydropower reservoirs in Scotland hold back some storage to store flood waters: if increased winter flows means an increased risk of flooding — as is likely — then more storage might need to be set aside for flood control. An increase in winter flows would not necessarily lead to an increase in hydropower generation potential in Scotland.

Thermal power generation

Thermal and nuclear power generation need water for cooling. In once-through cooling, water is abstracted from a river, lake or the sea, is circulated through the cooling system and is returned at considerably higher temperature. Closed-cycle cooling systems circulate the water through cooling towers. Considerably less is withdrawn — perhaps 5% of that needed for once-through cooling (Gleick, 1993) — but some of the water is lost through evaporation from the cooling towers. Abstractions for cooling water represent one of the largest water uses in many countries, including Britain.

There are two main effects of climate change. First, a rise in water temperature will reduce the efficiency of cooling so a greater volume of water may be demanded. Second, and more importantly, a change in river conditions will alter not only the amounts of water that can be abstracted, but also the amounts and temperatures at which water can be returned to watercourses. Several French nuclear power stations, for example, were forced to close down or operate well below design capacity during the European drought of 1991. Miller *et al.* (1993) examined the potential effects of global warming on the Tennessee Valley Authority (TVA) power supply system, and concluded that power generation would fall. This was due to plants being forced to run at lower levels, or even shut down for longer periods, because they would not meet regulatory standards for water temperature downstream: discharges would be added to warmer river flows, and thus temperature thresholds would be passed more frequently. Plant efficiency would be reduced and cooling towers would need to be used more often. Although the change in annual power production would be small, Miller *et al.* (1993) concluded that the system would be under greater pressure during extreme warm, dry periods.

In Britain, changes in river flows and water temperature would constrain abstractions for cooling, but over the next few decades there is likely to be a shift not only towards more efficient generation technologies, but also a move towards coastal locations.

Navigation

Rivers have long been important transport routes. In Europe, for example, both the Rhine and the Danube are the heart of major transport networks, and in North America the Mississippi is the major carrier of bulk goods.

River navigation is sensitive to climate change in four ways. First, a change in river flow variability would lead to a change in the opportunities for navigation and the cost of transporting goods. The maximum load of a vessel is determined by the maximum depth of water available. A reduction in flows (or lake water levels) would reduce water depths, reduce loads and so increase the number of trips required. Costs would therefore rise (as calculated by Marchand *et al.* (1988) for the Great Lakes). However the relationship between flow and cost is not linear and there is generally a threshold discharge below which loads must be reduced and transport costs rise. At the other end of the flow scale, high flows also constrain river navigation. As flow volumes and, more particularly, velocities rise, then the energy costs of moving upstream increase and navigation may be stopped. The effect of climate change on navigation would therefore depend on the frequency with which water levels pass these high and low thresholds.

The second potential effect of climate change is through changes in sedimentation within the river channel. In many major waterways, such as the Rhine, dredging may be necessary to maintain navigation. Alterations to sedimentation patterns are (as indicated in Chapter 7) very difficult to assess and may lead to changes in dredging costs. The navigation season in some regions is constrained by the duration of ice cover. With warmer temperatures, ice cover can be expected to reduce and the navigation season might be extended. The final, relatively minor, potential impact of climate change on navigation might arise through changes in the frequency of flooding of docking and other facilities, such as locks.

Many navigable waterways are managed through series of control reservoirs or instream structures, and these management structures might serve to mitigate some of the effects of climate change. Miller and Brock (1989), for example, found in the Tennessee valley that the reservoirs that currently sustain navigation would continue to do so in the future, under the scenarios considered.

Inland navigation in Britain

Although Britain has a long history of canal construction, dating back to the mid eighteenth century, the canals and rivers of Britain do not constitute a major part of the transport network. Most boats registered for use on Britain's waterways are private pleasure craft — nearly 92% in 1994 (British Waterways, 1995) — and most of the rest are canal hire boats. Only along a few short stretches of canal and navigable river, such as the lower Trent, some Yorkshire canals, the Manchester Ship Canal and the Gloucester and Sharpness Canal, are goods carried in large quantities, and the bulk of the canal and navigable river network is almost exclusively used for recreational boating.

The Environment Agency is responsible for managing navigation along a few rivers, including the Thames, Medway and lower Derwent and some rivers in East Anglia, but the entire canal network together with the remaining navigable rivers, including the Trent, are managed by British Waterways. Navigation in the navigable rivers is managed through the use of weirs and locks, and these exert considerable control over water levels along the river course. They can also store a certain amount of water to maintain levels during dry periods, but total storage volumes are small. The effects of climate change on navigation opportunities in such rivers will depend on how well the systems of weirs and locks can maintain levels under a scenario with increased flows in winter, and probably decreased flows during summer (Chapter 5). A second effect of climate change will be through altered patterns of erosion and, particularly, deposition, and it is likely that the costs of maintaining navigable waterways will alter.

The canal network — currently about 4000 km — takes water from rivers (where the two occasionally run in parallel) and from a large number of small, independent supply reservoirs. These are particularly prone to climatic variability and change. During dry years there may be severe shortages of water in the canal network, and parts of the network may be closed: 7.2% of the network suffered navigation restrictions during the dry summer of 1990 (British Waterways, 1991). Under the 1996 CCIRG scenario, it is likely that parts of the canal network, particularly in southern Britain, would be severely affected by water shortages, with possibly very large impacts on recreational boating and therefore large costs.

Recreation

Pleasure boating is one of the important recreational uses of the water environment. Other recreational uses include fishing — Britain's most popular sport — swimming, sailing, canoeing and less strenuous activities such as strolling along the riverside. Water-based recreation can be very important in economic and financial terms (Cooper, 1990), and many of the economic benefits of some American reservoirs have derived largely from the recreational opportunities they provide.

Climate change is likely to have an effect on the *demand* for water-based recreation and on *recreational potential*. However, it is very difficult to predict quantitatively the effects of change, so any assessment must be based on guesswork.

The demand for water-based recreation is obviously related to climate (Smith, 1990). It can be expected that the warmer and sunnier the climate, the greater the demand. Any effects of climate change, however, will be superimposed on background trends in recreational activity: the Environment Agency, for example, is already planning for large increases in visitors at its facilities.

It is slightly easier, though still difficult, to estimate changes in the potential for water-based recreation. The quantity of water is important for many uses, particularly boating, sailing and canoeing. For a given site it might be possible to predict changes in the frequency with which flows or levels remain within "suitable" limits (Ward, 1987; Brown et al., 1992). Water quality is also very important for water-based recreation (Burrows and House, 1989), and not only for activities involving contact with water: the aesthetic quality of the river environment may be strongly affected by the quality of the water. Changes in water quality, especially deterioration in water quality, are likely to alter the potential for water-based recreation. Algal bloom frequencies are particularly important

(Chapter 7). If they become more frequent, then health risks increase and aesthetic qualities decline. Fishing as a recreational activity would be affected not only by the aesthetic quality of the aquatic environment, but also by the number of fish that are caught. The next section looks briefly at the effects of climate change on aquatic ecosystems and fisheries, but it is worth noting here that Loomis and Lee (1993) estimated the lost recreational value due to reductions in salmon population along the Sacremento River, California, at around $35 million per year: this figure was calculated from the estimated fall in the number of fishing trips.

Aquatic ecosystems

In recent years it has increasingly been recognised that aquatic ecosystems count as instream uses of water, with legitimate demands on water resources. Aquatic ecosystems include those in the river channel or lake (instream ecosystems), and those in wetlands and riparian environments. All are dependent on the amount, quality and temperature of water, to varying degrees, and it is increasingly being realised that some aquatic ecosystems may be particularly sensitive to climate change (Firth and Fisher, 1992). The 1996 IPCC Assessment contains many examples of potential effects of climate change on freshwater ecosystems (Arnell *et al.*, 1996). However the major problem in estimating these potential effects is the poor current understanding of the links between climate and hydrology on the one hand, and the composition and dynamics of aquatic ecosystems on the other. There are few mechanistic models linking environmental controls with ecosystem properties (Poff, 1992), and many assessments of the effects of climate change are based either on empirical evidence from past climatic variability or on experimentation (for example by transferring animals or plants from one climate zone to another). Moreover, there is considerable uncertainty over the extent to which genetic variations between populations of the same species in different locations can influence adaptation to changed climates (Sweeney *et al.*, 1992).

The number of studies on freshwater ecosystems is rapidly increasing, and this brief review — distinguishing between instream ecosystems and wetland and riparian environments — summarises some of the main points.

Instream ecosystems

The species composition within a river is related to water temperature, hydrological regime, sediment characteristics and water quality. Within a lake, species composition is related more to temperature and thermal structure. Although a change in climate is likely to alter the instream environment, two dimensions of change are particularly important. The first is a change in the frequency of occurrence of extremes, either of temperature or flow. Disturbance during extreme events and subsequent patterns of recovery have a major impact on the characteristics of instream ecosystems (Poff and Ward, 1989), and changes in these disturbances will probably have a greater effect than changes in average conditions. Second, a change in the availability of *refugia* will have a great effect on species composition. Refugia are areas which provide critical habitat (such as spawning sites) or some other vital ecosystem function. They are important mediators of resistance to, and recovery from, disturbances (Sedell *et al.*, 1990) and they buffer against extreme hydrological events and other changes. The effect of climate change on instream ecosystems will depend on how the distribution, persistence and connectivity of refugia are affected. The link between "climate change" and "impact" may therefore be very complicated and may involve several important critical thresholds. In principle it is possible to simulate the effect of changes in flow regimes and water temperature on extreme disturbing events and refugia (it would simply involve calculating additional statistics from simulated flow series), but in practice this will be very difficult: the magnitude of "disturbance" and the dimensions of refugia vary from ecosystem to ecosystem (and species to species), and are very difficult to quantify.

Aquatic invertebrates are very sensitive to variations in stream or lake water temperature (Abdullahi, 1990; Moore *et al.*, 1995; Hogg *et al.*, 1995), during many stages of their life cycle. In general production rates increase with water temperature, but with global warming temperature may rise above thermal limits. For example a rise in autumn water temperature from 10°C to 16°C (which is slightly larger than might be anticipated under global warming) would be lethal to 99% of stonefly larvae. A rise in water temperature might also alter the timing of invertebrate growth stages: for example, eggs may hatch prematurely in autumn (Chen and Folt, 1996). Different invertebrate species are likely to be affected by change in temperatures in different ways, and species composition might therefore alter: this would have consequent effects for animals further up the food chain, including fish.

Fish populations will be influenced by changes in food sources, changes in water temperature and, perhaps more importantly, changes in river flows and sedimentation. Fish growth, survival and reproduction success are all related in some way to water temperature. Individual species have thermal limits, and between these limits physiological processes may be affected by temperature. Different fish species have different thermal requirements: in North America, fish species are divided into cold-water, cool-water and warm-water "guilds" (Magnuson*et al.*, 1979). Each guild would be differently affected by a rise in water temperature and, in general, cool-water and cold-water species would be most affected (Eaton and Scheller, 1996). They would lose habitat at the expense of warm-water species and geographic ranges should shift polewards (if new habitats are accessible through connecting waters). A rise in air temperature of 4°C, for example, would move the ranges of smallmouth bass and yellow perch in Canada northwards by up to 500 km (Shuter and Post, 1990). Minns and Moore (1993) simulated a northward displacement in the ranges of three species in Canadian lakes.

Hydrological regimes, particularly flow depth and velocity, also affect fish habitats and populations (Poff and Allan, 1995), as does sedimentation on the river bed. For many species there are hydrological limits (at the extreme, zero flow), but between these limits habitat suitability may vary with flow volume and the variability in flow over time. Despite the recent increase in research into links between flow, sediment and fish populations (Armitage, 1994), there have been very few attempts to explore the effects on fish habitat of changes in river flows resulting from climate change, with the notable exception of studies in semi-arid environments (Grimm and Fisher, 1992), where the interest focuses on changes in the frequency and duration of periods with no flow. Haneman and Dumas (1993), however, simulated a 50% reduction in chinook salmon populations in the Sacramento River, California, due largely to a reduced spawning habitat in autumn, increased scouring in winter and increased mortality in the Sacramento Delta. Changes in water quality can also be expected to lead to changes in fish populations. An increasing frequency of extreme events with low oxygen levels, for example, would increase stresses on fish populations (Covich, 1993).

Wetland and riparian environments

These environments are dominated by water, either through high water levels or by seasonal inundation. They are both very important ecologically and economically (as a source of food), and are very sensitive to change: this is shown by the many examples of loss of wetland due to various human activities such as drainage and abstraction of water.

A change in climate would affect wetland and riparian environments through changes in the hydrological regime as well as changes in air temperature. The key hydrological variables are water level within a wetland and the frequency and duration of inundation, both of which would be altered by global warming. These changes would lead to changes in species composition (Meyer and Pulliam, 1992), which might affect ecosystem productivity. Poiani and Johnson (1991) found that under the scenarios they considered, seasonal fluctuations in the volume of water in North American prairie wetlands would increase, and the lower summer levels would lead to reductions in important wildfowl habitats. Changes in hydrological regimes in wetlands might also affect biogeochemical processes which determine emissions of carbon dioxide and methane, and stream water quality (Cooper, 1990; Meyer and Pulliam, 1992). A fall in water levels, for example, would lead to increased decomposition of organic matter, with a consequent increase in discharge of nutrients to the stream.

The effects of climate change on aquatic ecosystems — both instream and wetland — are potentially very severe, but difficult to assess. Most work has been done on sensitivity of instream ecosystems and species to temperature change, and relatively little has been done on effects of changes in flows or on potential changes in wetland and riparian environments. There are very few mechanistic models of ecological processes which can simulate the effects of change, even though it is now feasible to provide scenarios of potential hydrological changes. What is clear, however, is that changes in extreme conditions — of temperature, drought or flooding — are likely to have much more effect on aquatic ecosystems than changes in average conditions, and that there are probably several important thresholds, beyond which substantial change will occur.

Aquatic ecosystems in Britain

The general principles outlined above apply, of course, to aquatic ecosystems in British rivers and lakes. Most British fish are well within their thermal limits and are unlikely to be significantly affected by changes in water temperature. Important exceptions are two cold-water lake fish: the whitefish (*Coregonus lavaretus*) and the charr (*Salvelinus alpinus*) — and two river fish: grayling (*Thymallus thymallus*) and native brown trout (*Salmo trutta*) (Arnell *et al.*, 1994). The two lake fish are currently close to their thermal limits, and their habitats would be very severely restricted by a rise in lake water temperature. The grayling could disappear from some southern rivers if water temperatures rise, and the growth rate of brown trout would decline: the non-native rainbow trout might have an increased competitive advantage. The decline of the brown trout in particular would have a significant impact upon fisheries. Native carp (*Cyprinus carpio*) and some exotic species — such as the zander (*Stizostedion lucioperca*) and grass carp (*Ctenopharyngodon idella*) — are currently restricted by low temperatures and their range might extend northwards in the future. Higher temperatures would alter the incidence of fish diseases and fungal infections, but some might increase whilst others decrease.

Changes in river flow regimes might be expected to have greater effects on British fish habitats and populations. An increasing frequency of summer droughts (Chapter 5), coupled with higher water temperatures and possibly lower dissolved oxygen contents, would pose increased threats to many fish populations. Elliot (1985), for example, showed how severe drought killed large numbers of brown trout in a small Cumbrian stream. High flows can also affect habitat, as eggs and fish are washed downstream and sediments are deposited over spawning beds. Finally, changes in river flows during the salmon migration season might alter migration rates, which are related to flow volumes and water temperature (Falkus, 1984; Scottish Fisheries Research Station, 1992). Increased flows would lead to increased migration rates, given enough salmon to sustain higher rates, but the water temperature window is between 4.4°C and 20°C (Falkus, 1984). Higher water temperatures would mean that salmon could enter rivers earlier, but might lay dormant sooner. Salmon migration patterns might also be affected by changes in currents and water temperature in the North Atlantic.

The management of aquatic weeds is a major maintenance issue in chalk streams in southern Britain. Increased water temperature, together with increased sunshine, can be expected to lead to an increase in weed growth, with consequent implications for river management. At present, weed growth tends to begin after the flood season has finished, and weeds have died back before the flood season begins again in late autumn. If weed growth is more persistent, increased maintenance might be necessary.

As in other parts of the world, there have been very few quantitative studies of the effect of changes in water level on wetlands in Britain, and it is therefore difficult to estimate impacts of climate change. It is important to note, however, that many important wetlands in Britain — the Norfolk Broads and Somerset Levels, for example — are largely sustained by management practices, and that these practices might mitigate, or at least complicate, the effects of climate change.

Competition and conflict

Water resources management is increasingly concerned with managing the competing demands of different users of water. This competition may be between similar types of users in different locations — upstream and downstream communities, for example — or between different uses — domestic, agricultural and environmental — in the same place. An altered pattern of streamflow has the potential to alter competition between users and locations, and where these users and locations are in different countries, there is scope for increased international conflict. Over the last few years, much has been written about conflict over scarce environmental resources, and much has focused around water (Thomas and Howlett, 1993).

More than 30 countries receive over one third of their water from upstream countries (Gleick, 1992). At the most extreme, Egypt and Hungary obtain 97% and 95% respectively of their water from upstream. There are currently many examples of tension over shared water resources — in the Middle East, the Nile basin, the Ganges basin and between Mexico and the United States, for example (McCaffrey, 1993) — and although other factors are also involved, a change in water resource availability could trigger changes in international relations. Many international rivers are covered by international treaties apportioning water, but none of these treaties can incorporate the uncertain effects of climate change (Gleick, 1988; 1992).

The allocation of scarce water amongst all the countries in an international basin provides one possible focus for international conflict, but another may arise from general inequalities in resource availability (Gleick, 1992). The nightmare scenario is of increased poverty in water-scarce regions triggering mass movement of environmental refugees towards water-rich areas (Tickell, 1993).

An altered distribution of resources can also lead to conflicts within a country (Frederick, 1993). Droughts have frequently led to disputes between regions (Stein, 1993), and during the drought of 1988-1992 there were serious disputes between Spanish regions over the transfer of water to support drought-stricken towns. Conflicts between users in the same basin are also likely to change, and if resources reduce, are most likely to increase. During the 1995 drought in Britain, for example, the National Rivers Authority opposed many applications to extract increased amounts of water from rivers to meet domestic demands, on the grounds that instream ecosystems and the aquatic environment had a higher priority.

Impacts on water resources: an overview

This section has reviewed the potential impacts of climate change upon different water uses, both internationally and in Britain. There are considerable uncertainties over the magnitude — and often direction — of impacts. Impacts on water supply and demand are the most studied and the best known: here, the major limitations lie in the credibility of the climate change scenarios used to generate changes in streamflow and groundwater recharge, and the incorporation of management responses to change. The models linking hydrological change to impact are relatively well-established. Several studies have looked at hydropower generation under a changed climate — but not in Britain — and again the models to translate hydrological change into impact are reasonably reliable. Impacts on thermal power generation have not really been investigated in much detail, and implications for navigation remain highly uncertain. The biggest area of uncertainty, however, lies in the estimation of the impacts of climate change on aquatic ecosystems. Here, the models linking change to impact are poorly developed, and even with credible scenarios for change in hydrology and water temperature it will be difficult to estimate implications for ecosystems and populations. The final part of this section introduced the possibility of altered conflict — between users, regions and countries — as a result of climate change. This possibility has had a high public profile in recent years, but it really is very difficult to infer changes in conflict from changes in water availability, when so many other factors contribute to national and international tension.

IMPACTS ON ENVIRONMENTAL RISK

There are essentially three main types of environmental risk or hazard associated with water: flood (too much water), drought (too little water) and disease and ill-health. To a large extent, droughts affect society through restrictions on water use, and as such have already been discussed, so the review here of climate change and droughts will be necessarily brief.

Floods

An increase in the frequency of flooding is among the most feared impacts of climate change, and is certainly one of the impacts with the highest public profile. However, as explained in Chapter 5, it is very difficult to assess changes in extreme high flows. This is largely because it is difficult to define credible scenarios for changes in the frequency of short-duration, flood-producing rainfall (or snowmelt) events, and it is significant that the most comprehensive study of the effect of climate change on flooding examined the Rhine River in Netherlands (Kwadijk and Middelkoop, 1994): floods here are an integration of snow and rainfall inputs over a period of several days or weeks, and short-duration events are not so important.

A change in flood regimes (Chapter 5) would have the effect of altering the frequency with which important flow thresholds are passed. Such thresholds include the level of overbank flooding, the top of a flood defence levee, the capacity of a flood storage pond, or the limits of a floodplain zone. Kwadijk and Middelkoop (1994) showed that the frequency of flows above

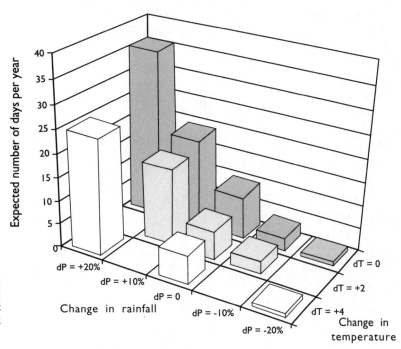

Figure 8.12 Number of days per year that peak flows exceed 5500 m³ s⁻¹ in the Rhine at Lobith, Netherlands (Kwadijk and Middelkoop, 1994)

5500 m³ s⁻¹ — sufficient to inundate most floodplains along the Rhine in the Netherlands — would alter dramatically if precipitation were to change (Figure 8.12).

Simple sensitivity analyses also indicate that a relatively small change in hydrological characteristics can produce large changes in flood risk. Figure 8.13 shows the return period of the current 100-year flood, under changes in the mean annual flood and the coefficient of variation of annual flood magnitude (Beran and Arnell, 1996). The figure was produced simply by altering the parameters of an Extreme Value Type 1 (or Gumbel) distribution, and calculating the return period under the new parameters of defined critical thresholds; the starting parameters of the distribution (mean = 10; CV = 0.4) are typical of properties of annual maximum floods in Britain. If both the mean annual flood and the CV of annual floods were to increase by 10%, which is consistent with climate change scenarios (Chapter 5), the flood currently exceeded on average once every 100 years would have a return period of approximately once in 40 years. Different frequency distributions would give different implied sensitivities to change, so Figure 8.13 should only be seen as an indication of the possible magnitude of change in risk.

A change in the frequency of flooding could have three major impacts on society. First, losses due to floods would change, depending on the change in flood frequency. Techniques are now well established to estimate the average annual flood damage at a site (Penning-Rowsell and Chatterton, 1977; Klaus

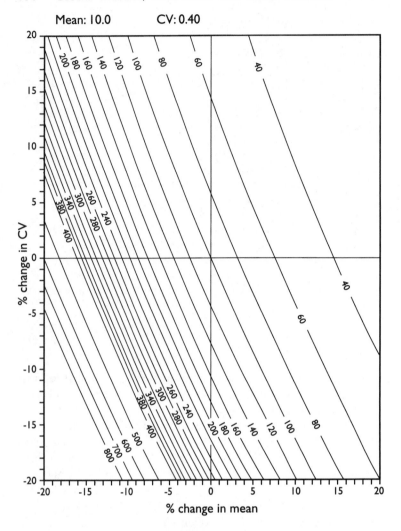

FIGURE 8.13 Return period of the current 100-year flood with changes in mean and coefficient of variation: EV I distribution assumed (Beran and Arnell, 1996)

et al., 1994), through combining information on the relationship between flood stage and flood damage, with the site stage-frequency relationship. The average annual damage is simply the area under the curve plotting flood damage against the probability of that damage being exceeded, and given a site stage-damage curve it would be relatively easy to determine the effects of changes in the site stage-frequency relationship. However, this has only been done by Smith (1993) for a site in Australia: he showed that by increasing the stage associated with a given frequency by an amount equal to one standard deviation of the annual stages, average annual damages would increase from $850 000 to over $7 million. This type of calculation indicates the extreme sensitivity of average annual damage to change in the frequency distribution. Average annual damage is largely determined by the frequency of the flood which just begins to cause damage, so changes in the frequency of this threshold will be particularly important. However, this sensitivity of estimates of average annual damage to changes in flood frequency does mean that estimates even under current climates can be very uncertain. Smith (1993), for example, showed that the estimated average annual damages at Lismore, Australia, varied by a factor of close to two if different periods of record were used, and this very large sampling uncertainty makes it difficult to assess the effects of climate change.

The second major implication of a change in flood risk is that existing standards of protection would alter. They might be increased, but the indications from Chapter 5 are that in most cases they would decrease. If standards were to decline, it would therefore be necessary either to accept the lower standard and bear the increased costs of flooding, or to invest in upgrading defences or management strategies to continue to meet existing standards. It may, however, be difficult and expensive to upgrade flood defences, for technical or political reasons. Smith (1993) points out that it would be politically very difficult to redefine floodplain zones in Australia.

The third potential impact of a change in flood frequency is through indirect effects on business productivity and, more importantly, the financial institutions of various types that pay for flood losses. In many countries governments pay for flood damages, either through loans, grants or tax relief, and an increase in government flood relief expenditure would obviously have to be met by reductions in other areas of government spending. In other countries, a proportion of flood losses is recovered through insurance. A rise in flood frequency would lead to increased payouts from the insurance industry, which would seek to recover these payouts through higher insurance and reinsurance premiums. Also, the insurance industry might seek to exclude some exposed properties from cover, thus affecting the economic viability of floodplain enterprises. At the most extreme, increasing insurance and reinsurance payouts could threaten the structure of the international insurance and finance market, simply by demanding too much cash (Dlugolecki *et al.*, 1995).

Changes in the flood risk in Britain

The implications of the simulations presented in Chapter 5 are that flood risk is unlikely to decline in Britain, and is most likely to increase. However, the magnitude of this increase is uncertain and has not been quantitatively assessed. The Environment Agency aims to provide urban communities with defined standards of service (where economically justifiable), and it is likely that the management response will be to invest to maintain these standards. The costs of this investment have not yet been determined, but may be very substantial.

An increasing number of planning authorities in Britain are implementing floodplain land use management policies, curbing new development in the floodplain. The most recent circular on floodplain development control, from the Department of the Environment, Ministry of Agriculture, Fisheries and Food, and the Welsh Office (1992) encourages floodplain land use control, and also reminds planning authorities of the potential threat posed by climate change. Climate change may therefore accelerate the adoption of floodplain planning, but other factors are likely to be more important.

Many flood-prone communities in Britain are covered by some form of flood warning system, often in combination with structural flood defences. Climate change might not affect the mechanics of issuing and disseminating a flood warning, but might mean that warnings need to be issued more frequently.

The insurance industry in Britain currently provides flood cover to domestic customers as part of the standard household policy. Riverine flooding is seen as a relatively minor problem by the insurance industry, which is devoting much more attention to coastal flooding and the potential effects of sea level rise. However, an increase in the frequency of riverine flooding, and hence flood claims, is likely to lead to increasing selection by the insurance industry, through differential premiums or excesses, or by refusal of cover. This selection is made increasingly feasible by technological advances, such as postcode premium rating, and could lead to declines in property values in high-risk floodplain areas.

Finally, the maintenance of flood defences and river channels is influenced by the frequency of occurrence of extreme damaging events, and also by patterns of erosion and deposition. Changes in erosion and deposition are unfortunately very difficult to assess (Chapter 7), so estimates of changed maintenance requirements are difficult to make. Weed growth too needs to be controlled in order to minimise flood risk in some channels, and as indicated above, changed temperature and sunshine might lead to increases in the duration of weed growth and hence in the need for weed clearance.

Health

There are an estimated 250 million additional cases of water-related disease each year, resulting in an annual total of approximately 10 million deaths (Nash, 1993). Water-related diseases and ill-health are conventionally divided into four groups. *Fecal-oral* diseases include cholera, diarrhoea, dysentery and typhoid. Water is the passive agent for the transmission of pathogens from excreta to humans. *Water-washed* diseases include infectious skin and eye diseases, and are caused by washing with infected water. *Water-based* diseases are transmitted by means other than washing or ingestion, by hosts that live in the water. Examples include schistosomiasis and guinea worm. *Water-related insect vector diseases* are transmitted by insects that spend all or part of their life in or near water. They include malaria, river blindness (onchocerciasis) and yellow fever. The first three types of illness are associated with poor water quality — either of drinking water or bathing water — whilst the last group are related more to the quantity of water, and particularly to the variation in water availability through the year.

To these microbiological diseases can be added those caused by excessive concentrations of chemicals, radiation or toxic metals, arising as a result of pollution. Examples of potentially dangerous substances — at least in excess — include heavy metals, organic pesticides and nitrates. These too are associated with water quality, rather than water quantity.

There are many factors triggering changes in water-related ill health (Nash, 1993). The most important is the combined effect of population growth, urbanisation and poverty, which are increasing the number of people with access to poor quality water. Industrial expansion and agricultural development affect the incidence of chemical-based illnesses, whilst large-scale water developments can provide increased opportunities for water-based and water-related insect vector diseases. The additional effects of climate change have not yet been determined, and are hard to ascertain.

The quality of drinking water, and hence the occurrence of fecal-oral and water-washed disease, may be affected by changes in water temperature, but the effects of poor sanitation are likely to be much more significant. The hosts that transmit water-based diseases generally live in standing water that is entered by humans for bathing or fishing. The life-cycle of these hosts may be affected by water temperature and the duration of standing water, but this is very uncertain. Water-related insect vector diseases are most likely to be affected by climate change, as a rise in temperature will tend to increase the geographic distribution of insect vectors, and a change in river flows or the duration of standing water will affect insect habitats. The tsetse fly, for example, is the vector for the parasite which causes river blindness (particularly in West Africa), and its life-cycle is closely associated with river flows and their variation through the year.

Water-related ill-health and disease are currently not a major problem in Britain, except during occasional pollution incidents. Particularly important are outbreaks of cryptosporidium in drinking water, which tend to be associated with effluent from pasture. The effects of climate change on such outbreaks and their intensity is not clear, but is likely to be minor. The other important potential health threat is through an increasing occurrence of toxic blue-green algal blooms on lakes, reservoirs and rivers. As indicated in Chapter 7, algal blooms may be affected by climate change, although other factors such as nutrient inputs are also important. Under global warming scenarios it is possible that malarial mosquitos might become re-established in Britain (CCIRG, 1996), but public health measures are likely to prevent significant outbreaks.

Droughts

The other high profile impact of climate change is on the frequency of occurrence of droughts. Drought, however, unlike flood, is difficult to define. A meteorological drought is a simple lack of rain. A climatic, or agricultural, drought is a lack of soil moisture, itself due to a lack of rain or high evaporation. A hydrological drought is a lack of river flow or groundwater recharge, whilst a water resources drought is a lack of water in a supply system. The duration of deficit is important, as is the intensity of deficit, but the baseline against which a deficit is measured, the critical intensity and the critical duration of deficit depend on the system of interest.

There have been several studies of climatological drought (e.g. Manabe and Wetherald, 1987; Mitchell and Warrilow, 1987; Rind et al., 1990), mostly using directly the output from global climate models. These have tended to show increased "drought" in mid-latitudes during summer, due to reduced rainfall and increased evaporation. A few studies have looked at soil moisture deficits using off-line water balance models (e.g. Whetton et al., 1993). Although there have been many studies into hydrological changes, few of these studies have looked explicitly at hydrological drought: most have concentrated on changes in mean flows (Chapter 5). Some studies have examined changes in low flow statistics, but drought durations and deficit volumes have not been calculated. It can be inferred that drought frequency would increase in those cases where average flows during the season of lowest flow decline. Few studies have looked at water resources droughts per se, although some have considered changes in the frequency of system "failure" (e.g. Wolock et al., 1993).

Droughts have an impact on the supply of water to all types of customers, but also have an effect on the aquatic environment. In many respects, droughts provide the biggest test for water management systems, as it is during droughts that competition for water is at its highest. There is a clear need for multi-sectoral assessments of the impacts of future droughts, as single-sector studies — focusing for example on water supply — only address part of the story. There have been several multi-sectoral studies of recent droughts in Europe and North America, partly aimed at drawing conclusions about possible future droughts (e.g. Cannell and Pitcairn, 1993; Dracup *et al.*, 1993). These studies all show how impacts and actions in one sector can affect the impact of drought in another.

Future droughts in Britain

The most serious droughts in Britain — such as 1976 and that lasting between 1988 and 1992 — tend to occur when dry winters are followed by warm summers. Demands are higher, and the resource base weaker because reservoirs and aquifers have not been fully replenished. There are exceptions, however, and the drought of 1995 in west Yorkshire followed not a dry winter, but a dry spring: the many small reservoirs supplying the region began to be emptied earlier in the year than is usual and were, unusually, not kept topped up by spring rainfall. In southern parts of Britain supplied by the chalk aquifer, there were few resource problems during the dry summer of 1995 because groundwater reserves had been replenished during the wet winter.

Water managers in Britain have several tools that can be applied during drought conditions. The first actions usually taken are to adopt measures to manage resources more effectively, and to encourage consumers to use less water. Source management measures include resting sources at most risk and construction of temporary pipelines. Under the 1991 Water Industry Act, water companies can impose a temporary ban on the use of garden sprinklers and hosepipes. The water regulator OFWAT has set a target for hosepipe bans of not more than once in ten years. Hosepipe bans can be in response to either total demand or peak demand, but are very unpopular with consumers.

The next stage is for a water company to apply for a Drought Order, either to authorise increased abstraction from new or existing sources, or to allow certain specified uses to be temporarily prohibited. Drought Orders are issued by the Secretary of State for the Environment, taking into account advice from the Environment Agency (who can apply for Drought Orders to alter the management of their own resource support schemes). Drought Orders authorising a company to limit or prohibit specified uses of water typically cover activities such as watering public parks, filling swimming pools, washing vehicles and cleaning buildings. Since 1995 the Environment Agency has been able to apply for Environmental Drought Orders, to protect instream ecosystems during drought conditions.

The final, most drastic, stage is for a water company to apply for an Emergency Drought Order, enabling it to impose rota cuts on customers and supply water only through standpipes and water tankers. The Secretary of State for the Environment, will only issue an Emergency Drought Order if satisfied that the deficiency of water is likely to impair the social and economic well-being of the population. OFWAT standards of service allow rota cuts or standpipes only once every 100 years.

The change in river flows in Britain outlined in Chapter 5 suggests that, at least in southern Britain, drought conditions would become more frequent and the above procedures would need to be applied more often. OFWAT, the Department of the Environment and the Environment Agency are all unlikely to be prepared to see reductions in the standard of service provided, either to water company customers or to the aquatic environment, so water companies will be expected to increase their efforts to cope with drought conditions. These efforts, which the water companies will want to be at the lowest cost, would probably take the form of improved efficiency, leakage control, improved links between sources and finally new sources.

IMPLICATIONS FOR WATER MANAGEMENT

The previous section has summarised some of the potential impacts of climate change on water resources, and has discussed at times possible adaptation to change. This section explores the implications of climate change for water management over the next few decades in more detail, focusing on three issues: possible adaptive strategies and factors influencing adaptability to change, the effects of uncertainty and other non-climatic pressures on adaptive strategies, and changes in techniques for water resources assessment.

Adaptive strategies

The impacts of climate change on water users and the water environment will depend on the nature of the institutions managing water and the physical infrastructure. In the most general terms, the more flexible and adaptable the institutional structure, the better able it will be to respond to climate change (Miller, 1989; Riebsame, 1988). Several factors influence the ability of a water management system to respond to change. At the highest level, these include financial resources, technical expertise and the need for change: Miller (1989) notes that institutional development can be driven by the increase in the marginal value of water. At a more detailed level, the legal framework may influence the ability of a system to adapt: for example, licences for abstraction may be granted for long periods, and may be altered only after compensation is paid.

The range of potential adaptive strategies is wide, and includes infrastructure developments, more efficient exploitation of existing resources, management of demand and reallocation of resources amongst different users according to altered priorities. Infrastructure developments, such as new reservoirs, however, may be very expensive and difficult to implement, on political and environmental grounds, and it is likely that future water resources management will focus on smaller-scale adjustments, demand management and resource reallocation (Riebsame, 1988). Water management response to increasing water shortages under current climatic conditions can provide an analogue for possible response in the future. Miller (1989) studied the management of irrigation and hydropower in the Snake River, Idaho, during a period of increasing competition, and found that management institutions evolved by clarifying water rights and implementing a water bank which allowed transfer between users. Rhodes *et al.* (1992) saw Denver's response to the veto of a proposed new supply reservoir as an analogue for future climate change, and found that the management system responded by renegotiating rights, encouraging water conservation and by attempting inter-basin transfers (although these have environmental costs).

Change in water management occurs in a series of small, incremental adjustments based on accumulated experience, and more drastic, crisis-oriented responses: Riebsame (1988) showed that this was how water management had evolved in the Sacremento basin, California. In Britain, the procedures for coping with droughts have evolved through experience during major droughts such as that of 1976. Riebsame (1988) notes that the mechanisms behind crisis-based adjustment are reasonably well known, but less is known about how gradual adjustment occurs. At present, operating rules and procedures are periodically revised when more information becomes available, but this revision tends to be based on past experience rather than forecasts of the future. It is also important to distinguish between adjustments which are specifically designed to cope with climate change (purposeful adjustments: Kates, 1985), and those which are primarily aimed at responding to some other pressure, such as increased demand or legislative changes (incidental adjustments). In practice, it is unlikely that many water management agencies will change their actions radically in the face of climate change, but they will adapt procedures already in place for coping with other changes. These procedures may serve to mitigate the effects of climate change without the need for any explicit action, but in some circumstances could conceivably make water resources more sensitive to climatic change. Stakhiv (1993) argues forcefully that the array of techniques already available to, and being implemented by, water managers can cope with climate change.

Throughout much of the developed world, the focus of water management is shifting away from large infrastructure schemes towards flexible, integrated systems incorporating both supply and demand management. Such systems are inherently better able to cope with climate change, and are easier to adjust as more information appears. In some parts of the water sector it will not be possible to adapt purposefully to change. The managed part — water supply, navigation, power generation, for example — should be able to develop strategies to cope with altered conditions, but it will be much harder to manage the impact of change on aquatic ecosystems. At the most basic, such ecosystems will adapt naturally to changed conditions, and will reach new equilibrium states which may have less, or more, value (in human terms) than at present. Human intervention can help natural ecosystems reach new stable conditions, perhaps by providing connections along a river corridor, and lack of intervention may prevent stable ecosystems developing.

The time scales over which the water industry operates vary considerably. Operational activities, such as river channel maintenance, have a short planning horizon and can be easily altered, while planning for major infrastructure works, such as a reservoir, can take 15 to 25 years (NRA, 1994) and design life may be over 50 years. The possibility of climate change needs to be factored in to these schemes, whilst the uncertainty in climate change effects needs to be considered explicitly.

Uncertainty and response to climate change

Schwartz and Dillard (1990) surveyed attitudes to climate change amongst water managers in a number of cities in the United States. Virtually none saw a need for immediate concern or action, but all were waiting for a scientific consensus to emerge. They saw other changes as more significant, particularly in the short and medium term. Stakhiv (1993) and Rogers (1994) emphasise how climate change is only one of the challenges facing water managers: "climate change is just one of several factors that make precise prediction impossible" (Rogers, 1994: p. 181). These other factors include demand growth, technological change, legislative change and change in water management objectives. All might change more rapidly than the resource base, and all are highly uncertain. Rogers (1994) noted that even a strong climate change trend would be small relative to year-to-year variability (see Chapter 6), and that over the medium term (say less than 40 years) management decisions would probably be no different than if there was no underlying trend.

James *et al.* (1969) examined the relative importance of uncertainties in hydrological characteristics, water quality changes, choice of planning goals and economic assumptions (specifically concerning demand for water) in estimating future water resources in the Potomac Basin. They concluded that economic and political variables were most important, and that hydrological uncertainty was least important: Rogers (1994) drew a cautionary lesson for climate change assessments and their exaggeration (in his view) of the effects of climate change.

So should water planners ignore climate change, and wait either for stronger evidence to emerge or a consensus to form about the local magnitude of future change? Given the uncertain nature of change, and the small size of trend relative to year-to-year variability, it is probably appropriate that no explicit action needs to be taken for schemes with combined planning periods and operational lives under 10 years: there will be enough time to react to better climate change information as it becomes available. For schemes and plans looking longer into the future, however — particularly reservoir schemes which may be around for up to 100 years — it would be imprudent not to consider the additional effects of climate change *alongside the effects of other changes*. Such an assessment would give an indication of the potential magnitude of climate change effects, and whether the proposed scheme would be significantly affected. Indeed, robustness to an uncertain future may be seen as a design criterion: "The best projects are those that make sense even when the future turns out worse than forecast." (Kelly and Kelly, 1986: p.115).

The Environment Agency, for example, is aware of the threat of climate change, but does not account for it explicitly in planning or constructing water management schemes. However, in its water resource management role it has adopted the precautionary principle — "Where significant environmental damage may occur, but knowledge on the matter is incomplete, decisions made and measures implemented should err on the side of caution." (NRA, 1994: p. 2). The Environment Agency has also begun to build in scope for enhancing coastal flood defences in the future if sea level rise occurs (Arnell *et al.*, 1994).

Estimating future conditions

Current engineering practice is to assume that data from the past can be used to estimate the hydrological conditions of the future. Standard hydrological techniques, such as those in the UK Low Flows Studies Report (Gustard *et al.*, 1992) and Flood Studies Report (NERC, 1975) assume a stable climate. If climate change is occurring, then these assumptions cannot hold, and some means must be found to account for uncertain non-stationarity in hydrological time series. There are several key issues, which need methodological investigation. First, how is it possible to index hydrological behaviour against a changing background? Risk assessments might need to be time specific (for example, stating that there is a 1 in 10 chance of an event greater than some threshold over the first ten years of scheme life, and a 1 in 7 chance in the next ten). Second, how is it possible to add information on future climate trends to hydrological data to assess future risk?

The obvious solution would be to use simulated future hydrological data, but this would require not only credible change scenarios, but also very credible hydrological models. The major conceptual disadvantage of this approach is that it would base design on hypothetical, un-validated data, rather than — as is generally the case at present — on observed experience (although a large number of schemes are currently designed using synthetic data where observed data are unavailable).

The third key issue is that of uncertainty in climate change forecasts and the best ways of incorporating it into design estimates. Techniques already exist for coping with statistical error (through the use of safety factors and confidence intervals), but climate change is likely to add considerable extra uncertainty. Bayesian techniques are a possibility, but in order to be applied most effectively these would need credible assessments of the likelihood of different alternative futures: this is very difficult to determine.

SOME CONCLUDING COMMENTS

This book has explored some potential effects of climate change on hydrological characteristics and water resources in Britain, bringing in examples from other countries. Much of the book has been devoted to methodological issues, and the actual assessment of change is largely based on just one climate change scenario (CCIRG, 1996). Different scenarios would give different magnitudes of change, but it is likely that the general pattern of change would be similar, and all change is a potential threat to a system and its managers.

There are some major gaps in understanding of the potential effects of climate change on water resources, not just in Britain but worldwide. The most important gaps are:

● *Credible climate change scenarios at the catchment scale*. This requires credible climate simulation at global and regional scales, but also the use of reliable techniques for downscaling climate model simulations to the catchment scale. The use of nested regional climate models appears to offer many benefits, since it not only leads to more reliable estimates of local climate, but also gives information on changes in short-term climatic characteristics at hydrologically-relevant spatial scales. However, the effort involved in embedding regional models may not be justified by the credibility of the global-scale simulations. Scenarios need to consider change not only in mean climate, but also in the variability.

- *Robust models to translate local climate changes into biophysical impacts.* Several areas of research are necessary. First, there is a need to use integrated hydrological and atmospheric models, so that the simulated changes in hydrological conditions are fed back into the atmospheric model in a more accurate way than at present. However, for studies at the catchment scale (say less than 1000 km^2) there will be few advantages in using a coupled hydrological-atmospheric model over an off-line hydrological model fed with the output from a climate model. Second, there is a need to improve the simulation of important processes within the hydrological cycle, and particularly the generation of flood flows. Third, process-based models of water quality need to be improved and applied in climate change studies: possible changes in water quality are very uncertain. Fourth, ecological models translating environmental change into ecological impact (at the organism, population and community scales) need to be developed and applied. As indicated above, the potential effects of climate change on aquatic ecosystems are very large, but are presently poorly understood.
- *Studies to estimate the impacts of these changes, in economic and financial terms, in different parts of the water sector.* Most "impact" studies have so far concentrated on water supply, and few have looked at the economic and financial consequences. Studies are needed in other impact areas, such as navigation and power generation. Studies of the impact of change, however, are closely linked to:
- *Studies into adaptation to change.* Possible adaptive strategies need to be explored, and barriers to efficient adaptation investigated. Particularly important are studies into decision-making in an uncertain environment.
- *Identification of critical sensitivities.* Virtually all climate change impact studies have explored the effects of a given change in climate. It can be very illuminating to look at the problem in the reverse direction and identify the type or degree of change that would be necessary to have some defined impact upon a system.

This book represents an early stage in the assessment of the potential impacts of climate change on water resources. It has taken stock of the current state of knowledge, and also suggested directions for future research into this vitally important issue.

Acknowledgements

Many colleagues at the Institute of Hydrology contributed, directly and indirectly, to the research reviewed in this book. Nick Reynard played a particularly important role through his construction of climate change scenarios, and Max Beran, Professor Brian Wilkinson and Frank Law provided helpful advice throughout. The water quality studies reviewed in Chapter 7 were led by Dr Alan Jenkins. More generally, colleagues throughout the Institute provided the ideal intellectual environment for research in water science: many ideas and insights sprang from coffee-time discussions.

The book has also benefited significantly from help and advice from many people outside the Institute. Dr Mike Hulme (Climatic Research Unit, University of East Anglia) provided the baseline climate data, the long time series of temperature and precipitation introduced in Chapter 2, and the climate change scenarios. Colin Wright (Department of the Environment, Water Directorate), Mark Sitton and Dr Richard Streeter (National Rivers Authority) all gave very helpful information on water management in Britain, and the sections on demand for water benefited greatly from discussions with Paul Herrington (University of Leicester). Many of the simulations of future British water resources described in the book were undertaken as a contribution to the work of the UK Climate Change Impacts Review Group (CCIRG), chaired by Professor Martin Parry (University College London). Colleagues in the Intergovernmental Panel on Climate Change (IPCC) working groups on impacts on hydrology and water resources also provided a sounding board for ideas and concepts - some of which got through to the book.Some of the work presented in this book draws on research funded by the UK Department of the Environment Water Directorate (PECD/7/7/348), and by the National Rivers Authority.

The book was produced by Charlotte Allen, Nick Fey, John Griffin and Celia Kirby of the Publications Section at the Institute of Hydrology.

Finally, I would like to thank Hilary for her help and encouragement throughout the production of this book, and I hope — for the sake of James, Fay and Rosie — that the worst impacts of global warming described in this book don't actually happen.

References

Abbott, M.B., Bathurst, J.C., Cunge, J.A., O'Connell, P.E. & Rasmussen, J. 1986. An introduction to the European Hydrological System — Système Hydrologique Européen, SHE. 1. History and philosophy of a physically-based, distributed modelling system. *J. Hydrol.* **87**, 45-59.

Abdullahi, B.A. 1990. The effect of temperature on reproduction in three species of cyclopoid copepods. *Hydrobiol.* **196**, 101-109.

Allen, R.G., Gichuki, F.N. & Rosenzweig, C. 1991. CO_2-induced climatic changes in irrigation water requirements. *J. Water Res. Plan. Manage.* **117**, 157-178.

Andre, J.-C., Goutourbe, J.P., Perrier, A. *et al.* 1988. Evaporation over land surfaces: first results from HAPEX-MOBILHY special observing period. *Ann. Geophys.* **6**, 477-492.

Armitage, P.D. 1994. Prediction of biological responses. In: Calow, P. & Petts, G.E. (Eds.) *The Rivers Handbook.* Vol. 2. Blackwell, Oxford. 254-275.

Arnell, N.W. 1992. Factors controlling the effects of climate change on river flow regimes in a humid temperate environment. *J. Hydrol.* **132**, 321-342.

Arnell, N.W. 1994. Hydrology and climate change. In: Calow, P. & Petts, G.E. (Eds.) *The Rivers Handbook.* Vol. 2. Blackwell, Oxford. 173-186.

Arnell, N.W. 1995a. Socioeconomic impacts of changes in water resources due to global warming. In: Oliver, H. & Oliver, S.A. (Eds.) *The Role of Water and the Hydrological Cycle in Global Change.* NATO ASI Series. Series 1: Global Environmental Change, Vol. 31. Springer, Berlin. 429-457.

Arnell, N.W. 1995b. Scenarios for hydrological climate change impact studies. In: Oliver, H. & Oliver, S.A. (Eds.) *The Role of Water and the Hydrological Cycle in Global Change.* NATO ASI Series. Series 1: Global Environmental Change, Vol. 31. Springer, Berlin. 389-407.

Arnell, N.W. 1996. Palæohydrology and future climate change. In: Branson, J., Brown, A.G. & Gregory, K.J. (Eds.) *Global Continental Changes: the Context of Palæohydrology.* Geol. Soc. Lond. Special Publication No.115, 19-25.

Arnell, N.W. & Dubourg, W.R. 1995. Implications for water supply and water management. In: Parry, M.L. & Duncan, R. (Eds.) *The Economic Implications of Climate Change for Britain.* Earthscan, London. 28-45.

Arnell, N.W. & Piper, B.S. 1996. Impact of climate change on an irrigation scheme in Lesotho. In: Kaczmarek, Z. *et al.* (Eds.), *Water Resources Management in the Face of Climatic and Hydrologic Uncertainties.* Kluwer, Dordrecht. *in press.*

Arnell, N.W. & Reynard, N.S. 1989. Estimating the impacts of climate change on river flows: some examples from Britain. *Conference on Climate and Water.* Helsinki, September 1989. Publications of the Academy of Finland. Vol. 1, 426-436.

Arnell, N.W. & Reynard, N.S. 1993. *Impact of Climate Change on River Flow Regimes in the United Kingdom.* Institute of Hydrology. Report to the Department of the Environment. 130pp.

Arnell, N.W. & Reynard, N.S. 1996. The effects of climate change due to global warming on river flows in Great Britain *J. Hydrol. in press.*

Arnell, N.W., Brown, R.P.C. & Reynard, N.S. 1990. *Impact of Climatic Variability and Change on River Flow Regimes in the UK.* Institute of Hydrology, Report No. 107. Wallingford, Oxon, UK.

Arnell, N.W., Jenkins, A. & George, D.G. 1994. *The Implications of Climate Change for the National Rivers Authority.* NRA R&D Report 12. 94pp.

Arnell, N.W., Bates, B., Lang, H., Magnuson, J.J. & Mulholland, P. 1996. Hydrology and Freshwater Ecology. In: *Climate Change 1995: Impacts, Adaptations and Mitigation of Climate Change. Scientific-Technical Analyses.* Contribution of Working Group II to the Second Assessment Report of the Intergovernmental Panel on Climate Change. Watson, R.T., Zinyowera, M.C. & Moss, R.H. (Eds.) Cambridge University Press, Cambridge and New York. 325-363.

Bardossy, A. & Caspary, H.J. 1991. Conceptual model for the calculation of the regional hydrologic effects of climate change. In: *Hydrology for the Water Management of Large River Basins.* Int. Assn. Hydrol. Sci. Publ. **201**, 73-83.

Bardossy, A. & Plate, E.F. 1992. Space-time model for daily rainfall using atmospheric circulation patterns. *Wat. Resour. Res.* **28**, 1247-1259.

Bates, B.C., Charles, S.P., Sumner, N.R. & Fleming, P.M. 1994. Climate change and its hydrological implications for South Australia.*Trans. Roy. Soc. of Sci. Aust.* **118**, 35-43.

Beran, M.A. & Arnell, N.W. 1996. Climate change and hydrological disasters. In: Singh, V.P. (Ed.), *Hydrology of Disasters.* Kluwer, Dordrecht, 41-62.

Blaney, H.F. & Criddle, W.D. 1950. Determining water requirements in irrigated areas from climatological and irrigation data. *USDA (ARS) Technical Bulletin* **1275**. 59pp.

Blumberg, A.F. & DiToro, D.M. 1990. Effects of climate warming on dissolved oxygen concentrations in Lake Erie. *Trans. Am. Fisheries Soc.* **119**, 210-223.

Boardman, J., Evans, R., Favis-Mortlock, D.T. & Harris, T.M. 1990. Climate change and soil erosion on agricultural land in England and Wales. *Land Degrad. Rehab.* **2**, 95-106.

British Waterways. 1991. *Annual Report and Accounts, 1991.* British Waterways, Watford.

British Waterways. 1995. *Report and Accounts, 1994/95.* British Waterways, Watford.

Broecker, W.S., Peteet, D. & Rind, D. 1985. Does the ocean-atmosphere system have more than one stable mode of operation? *Nature* **328**, 123-126.

Brown, B.G. & Katz, R.W. 1995 Regional analysis of temperature extremes: spatial analog for climate change? *J. Climate* **8**, 108-119.

Brown, T.C., Taylor, J.G. & Shelby, B. 1992. Assessing the direct effects of streamflow on recreation: a literature review. *Wat. Resour. Bull.* **27**, 979-989.

Budyko, M.I. 1982 *The Earth's Climate: Past and Future.* International Geophysics Series. Vol. 29. Academic Press: New York.

Bultot, F., Coppens, A., Dupriez, G.L., Gellens, D. & Meulenberghs, F. 1988a. Repercussions of a CO_2-doubling on the water cycle and on the water balance — a case study for Belgium.*J. Hydrol.* **99**, 319-347.

Bultot, F., Dupriez, G.L. & Gellens, D. 1988b. Estimated annual regime of energy balance components, evapotranspiration and soil moisture for a drainage basin in the case of CO_2 doubling. *Climatic Change* **12**, 39-56.

Bultot, F., Gellens, D., Spreafico, M. & Schädler, B. 1992. Repercussions of a CO_2-doubling on the water balance — a case study in Switzerland. *J. Hydrol.* **137**, 199-208.

Burn, D.H. 1994. Hydrologic effects of climatic change in west-central Canada. *J. Hydrol.* **160**, 53-70.

Burrows, A.M. & House, M.A. 1989. Public's perception of water quality and the use of water for recreation. In: Laikari, H. (Ed.) *River Basin Management*. Pergamon, Oxford. 371-379.

CCIRG (Climate Change Impacts Review Group). 1991. *The Potential Effects of Climate Change in the United Kingdom*. HMSO, London.

CCIRG (Climate Change Impacts Review Group) 1996. *Review of the Potential Effects of Climate Change in the United Kingdom*. HMSO, London.

Cannell, M.G.R. & Pitcairn, C.E.R. (Eds.) 1993. *Impacts of the Mild Winters and Hot Summers in the United Kingdom in 1988-1990*. HMSO, London.

Carter, T.R., Parry, M.L., Nishioka, S. & Harasawa, H. 1994. *Technical Guidelines for Assessing Climate Change Impacts and Adaptations*. Intergovernmental Panel on Climate Change, University College London, and Center for Global Environmental Research, Tsukuba. 51pp.

Chen, C.Y. & Folt, C.L. 1996. Consequences of fall warming for zooplankton overwintering success. *Limnol. Oceanogr.* **41**, *in press.*

Chiew, F.H.S. & McMahon, T.A. 1993. Detection of trend or change in annual flow of Australian rivers. *Int. J. Climatol.* **13**, 643-653.

Chiew, F.H.S, Whetton, P.H., McMahon, T.A. & Pittock, A.B. 1995 Simulation of the impacts of climate change on runoff and soil moisture in Australian catchments. *J. Hydrol.* **167**, 121-147.

Clark, K.J., Clark, L., Cole, J.A., Slade, S. & Spoel, N. 1992. *Effect of Sea Level Rise on Water Resources*. WRc plc. National Rivers Authority R&D Note 74.

Cohen, S.J. 1986. Impacts of a CO_2-induced climatic change on water resources in the Great Lakes basin. *Climatic Change* **8**, 135-153.

Cohen, S.J. 1987a. Sensitivity of water resources in the Great Lakes region to changes in temperature, precipitation, humidity and windspeed. In: *The Influence of Climate Change and Climatic Variability on the Hydrologic Regime and Water Resources*. Int. Assn Hydrol. Sci. Publ. **168**, 489-500.

Cohen, S.J. 1987b. Projected increases in municipal water use in the Great Lakes due to CO_2-induced climatic change. *Wat. Resour. Bull.* **23**, 81-101.

Cohen, S.J. 1991. Possible impacts of climatic warming scenarios on water resources in the Saskatchewan River sub-basin, Canada. *Climatic Change* **19**, 291-317.

Cohen, S.J. 1995. *McKenzie Basin Impact Study*. Summary of Interim Report #2. Climate Change Digest CCD 95-01. Environment Canada, Downsview, Ontario.

Cole, J.A., Slade, S., Jones, P.D. & Gregory, J.M. 1991. Reliable yield of reservoirs and possible effects of climatic change. *Hydrol. Sci. J.* **36**, 579-597.

Cooper, C.F. 1990. Recreation and wildlife. In: Waggoner, P.E. (Ed.) *Climate Change and US Water Resources*. Wiley, New York. 329-339.

Cooper, D.M., Wilkinson, W.B. & Arnell, N.W. 1995. The effect of climate change on aquifer storage and river baseflow. *Hydrol. Sci. J.* **40**, 615-631.

Cooter, E.J. & Cooter, W.S. 1990. Impacts of greenhouse warming on water temperature and water quality in the southern United States. *Climate Research* **1**, 1-12.

Covich, A.P. 1993. Water and ecosystems. In: Gleick, P.H. (Ed.) *Water in Crisis*. Oxford University Press, New York. 40-55.

Dansgaard, W., White, J.W.C. & Johnsen, S.J. 1989. The abrupt termination of the Younger Dryas climate event. *Nature* **339**, 532-533.

Department of the Environment. 1992.*Using Water Wisely*. A Consultation Paper. Department of the Environment and Welsh Office. 36pp.

Department of the Environment, Ministry of Agriculture, Fisheries and Food, and Welsh Office. 1992. *Development in Flood Risk Areas*. Circular 30/92.

Department of Trade and Industry 1995. Digest of Energy Statistics 1995. London: HMSO.

Dlugolecki, A., Harrison, P., Leggett, J. & Palutikof, J. 1995. Implications for insurance and finance. In: Parry, M.L. & Duncan, R. (Eds.) *The Economic Implications of Climate Change for Britain*. Earthscan: London. 83-102.

Dracup, J.A., Pelmuder, S.D., Howitt, R. *et al.* (Eds.) 1993. *Integrated Modelling of Drought and Global Warming: Impacts on Selected California Resources*. National Institute for Global Environmental Change, University of California, Davis. Report to US Environmental Protection Agency.

Drake, B.G. 1992. The impact of rising CO_2 on ecosystem production. *Water, Air and Soil Pollution* **64**, 25-44.

Dümenil, L. & Todini, E. 1992. A rainfall-runoff scheme for use in the Hamburg climate model. In: O'Kane, J.P. (Ed.)*Advances in Theoretical Hydrology*. European Geophysical Society Hydrological Science Series 1. Elsevier, Amsterdam. 129-157.

Eaton, J.G. & Scheller, R.M. 1996. Effects of climate warming on fish thermal habitat in streams of the United States. *Limnol. Oceanogr.* **41**, in press.

Elliot, J.M. 1985. Population dynamics of migratory trout *Salmo trutta* in a Lake District stream 1966-1983 and their implications for fishery management. *J. Fish Biol.* **27**, 35-43.

Erickson, D.J., Oglesby, R.J. & Marshall, S. 1995. Climate response to indirect anthropogenic sulphate forcing. *Geophys. Res. Lett.* **22**, 2017-2020.

Falkus, H. 1984. *Salmon Fishing*. Witherby, London.

Firth, P. & Fisher, S.G. (Eds.) 1992.*Global Climate Change and Freshwater Ecosystems*. Springer, Berlin.

Frederick, K.D. 1993. Climate change impacts on water resources and possible responses in the MINK region. *Climatic Change* **24**, 83-115.

Gellens, D. 1991. Impact of a CO_2-induced climate change on river flow variability in three rivers in Belgium. *Earth Surf. Processes & Landforms* **16**, 619-625.

Gellens, D. 1993. Sensitivity study of the hydrological cycle: impact of the climate change induced by the doubling of CO_2 atmospheric concentration. *Global Change Symposium*, Brussels, 17-18 May 1993. Science Policy Office, Belgium.

George, D.G. 1989. The thermal characteristics of lakes as a measure of climate change. *Conference on Climate and Water*. Helsinki, Sept. 1989. Publications of the Academy of Finland. Vol. 1, 402-412.

George, D.G., Hewitt, D.P., Lund, J.W.G. & Smyly, W.J.P. 1990. The relative effects of enrichment and climate change on the long-term dynamics of *Daphnia* in Esthwaite Water, Cumbria. *Freshwater Biol.* **23**, 55-70.

Ghassemi, F., Jacobson, G. & Jakeman, A.J. 1991. Major Australian aquifers: potential climate change impacts. *Water International* **16**, 38-44.

Gifford, R.M. 1988. Direct effect of higher carbon dioxide concentrations on vegetation. In: Pearman, G.L. (Ed.), *Greenhouse: Planning for Climatic Change*. CSIRO, Melbourne. 506-519.

Giorgi, F. & Bates, G.T. 1989. The climatological skill of a regional model over complex terrain. *Mon. Weath. Rev.* **117**, 2325-2347.

Giorgi, F. & Mearns, L.O. 1991. Approaches to the simulation of regional climate change: a review. *Rev. Geophys.* **29**, 191-216.

Giorgi, F., Marinucci, M.R. & Visconti, G. 1992. A $2\times CO_2$ climate change scenario over Europe generated using a limited area model nested in a general circulation model. II. Climate change scenario. *J. Geophys. Res.* **97**, D10, 10011-10028.

Giorgi, F., Brodeur, C.S. & Bates, G.T. 1994. Regional climate change scenarios over the United States produced with a nested regional climate model. *J. Climate* **7**, 375-399.

Gleick, P.H. 1986. Methods for evaluating the regional hydrologic impacts of global climatic changes. *J. Hydrol.* **88**, 97-116.

Gleick, P.H. 1987. Regional hydrologic consequences of increases in atmospheric CO_2 and other trace gases. *Climatic Change* **10**, 137-161.

Gleick, P.H. 1988. The effects of future climatic change on international water resources: the Colorado River, the United States and Mexico. *Policy Sciences* **21**, 23-39.

Gleick, P.H. 1992. Effects of climate change on shared water resources. In: Mintzer, I.M. (Ed.) *Confronting Climate Change: Risks, Implications and Responses*. Cambridge University Press, Cambridge. 127-140.

Gleick, P.H. 1993. Water and energy. In: Gleick, P.H. (Ed.) *Water in Crisis*. Oxford University Press, New York. 67-79.

Goutorbe, J., Lebel, T., Tinga, A. *et al.* 1993. HAPEX-Sahel: a large-scale study of land-atmosphere interactions in the semi-arid tropics. *Ann. Geophys.* **12**, 53-64.

Grimm, N.B. & Fisher, S.G. 1992. Responses of arid-land streams to changing climate. In: Firth, P. & Fisher, S.G. (Eds.) *Global Climate Change and Freshwater Ecosystems*. Springer, Berlin. 211-233.

Groisman, P.Ya. & Legates, D.R. 1995. Documenting and detecting long-term precipitation trends: where we are and what should be done. *Climatic Change* **31**, 601-622.

Gustard, A., Bullock, A. & Dixon, J. 1992. *Low Flow Estimation in the United Kingdom*. Institute of Hydrology Report No. 108. 88pp + app.

Haneman, W.M. & Dumas, C.F. 1993. Simulating impacts on Sacramento River fall run chinook salmon. In: Dracup, J.A., Pelmuder, S.D., Howitt, R. *et al.* (Eds.), *Integrated Modelling of Drought and Global Warming: Impacts on Selected California Resources*. National Institute for Global Environmental Change, University of California, Davis. Report to US Environmental Protection Agency. 69-95.

Haneman, W.M. & McCann, R. 1993. Economic impacts on the Northern California hydropower system. In: Dracup, J.A., Pelmuder, S.D., Howitt, R. *et al.* (Eds.) *Integrated Modelling of Drought and Global Warming: Impacts on Selected California Resources*. National Institute for Global Environmental Change, University of California, Davis. Report to US Environmental Protection Agency. 55-68.

Harding, R.J. & Moore, R.J. 1988. *Assessment of Snowmelt Models for Use in the Severn-Trent Flood Forecasting System*. Institute of Hydrology, Wallingford. 41pp.

Hassan, F.A. 1981. Historical Nile floods and their implications for climatic change. *Science* **212**, 1142-1145.

Hay, L.M., McCabe, G.J., Wolock, D.M. & Ayers, M.A. 1991. Simulation of precipitation by weather type analysis. *Wat. Resour. Res.* **27**, 493-501.

Hay, L.M., McCabe, G.J., Wolock, D.M. & Ayers, M.A. 1992. Use of weather types to disaggregate general circulation model predictions. *J. Geophys. Res.* **97**, D3, 2781-2790.

Hendry, G.R., Lewin, K.F. & Nagy, J. 1993. Free air carbon dioxide enrichment: development, progress, results. *Vegetatio* **104/105**, 17-31.

Herrington, P. 1996. *Climate Change and the Demand for Water.* HMSO, London. 164pp.

Hewett, B.A.O., Harries, C.D. & Fenn, C.R. 1993. Water resource planning in the uncertainty of climate change: a water company perspective. In: White, R. (Ed.), *Engineering for Climate Change.* Thomas Telford, London. 38-54.

Hisdal, H., Erup, J., Gudmundsson, K. *et al.* 1993. *Historical runoff variations in the Nordic countries.* Nordic Hydrological Programme Report 37. Nordic Coordinating Committee for Hydrology, Oslo. 100pp.

Hogg, I.D., Williams, D.D., Eadie, J.M. & Butt, S.A. 1995. The consequences of global warming for stream invertebrates: a field simulation. *J. Thermal Biol.* **20**, 199-206.

Hostetler, S.W. & Giorgi, F. 1993. Use of output from high-resolution atmospheric models in landscape-scale hydrologic models: an assessment. *Wat. Resour. Res.* **29**, 1685-1695.

Hulme, M. 1994a. Historic records and recent climatic change. In: Roberts, N. (Ed.) *The Changing Global Environment.* Blackwell, Oxford. 69-98.

Hulme, M. 1994b. Validation of large-scale precipitation fields in General Circulation Models. In: Desbois, M. & Desalmand, F. (Eds.), *Global Precipitation and Climate Change.* Springer-Verlag, Berlin. 387-405.

Hulme, M. 1996. *The 1996 CCIRG Scenario of Changing Climate and Sea Level for the United Kingdom.* Climate Impacts LINK Technical Note 7. Climatic Research Unit, University of East Anglia, Norwich. 44pp.

Hulme, M., Conway, D., Brown, O. & Barrow, E.M. 1994. *A 1961-90 Baseline Climatology and Future Climate Change Scenarios for Great Britain and Europe: Part III, Climate Change Scenarios for Great Britain and Europe.* Climatic Research Unit, University of East Anglia, Norwich. 29pp.

Hulme, M., Jiang, T. & Wigley, T.M.L. 1995a. *SCENGEN: a Climate Change Scenario Generator.* Climatic Research Unit. University of East Anglia, Norwich. 38pp.

Hulme, M., Raper, S.C.B. & Wigley, T.M.L. 1995b. An integrated framework to address climate change (ESCAPE) and further developments of the global and regional climate modules (MAGICC). *Energy Policy* **23**, 347-355.

Hulme, M., Conway, D., Jones, P.D. *et al.* 1995c. Construction of a 1961-1990 European climatology for climate change modelling and impact applications. *Int. J. Climatol.* 15, 1333-1363.

Imbrie, J. & Imbrie, K.P. 1979. *Ice Ages: Solving the Mystery.* Macmillan, London.

Institute of Hydrology. 1980. *Low Flow Studies.* Institute of Hydrology, Wallingford.

Institute of Hydrology. 1990. *Maidenhead, Windsor and Eton Flood Alleviation Scheme: Water Quality Study.* Institute of Hydrology, Wallingford. Report to National Rivers Authority Thames Region.

IPCC (Intergovernmental Panel on Climate Change) 1990a. *Climate Change. The IPCC Scientific Assessment.* Houghton, J.T., Jenkins, G.J. & Ephraums, J.J. (Eds.), Cambridge University Press: Cambridge.

IPCC (Intergovernmental Panel on Climate Change). 1990b. Climate *Change. The IPCC Impacts Assessment.* Tegart, W.J.McG., Sheldon, G.W. & Griffiths, D.C. (Eds.), Australian Government Publishing Service, Canberra.

IPCC (Intergovernmental Panel on Climate Change). 1991. *Climate Change. The IPCC Response Strategies.* Island Press, Washington D.C.

IPCC (Intergovernmental Panel on Climate Change). 1992. *Climate Change 1992. The Supplementary Report to the IPCC Scientific Assessment.* Houghton, J.T., Callendar, B.A. & Varney, S.K. (Eds.), Cambridge University Press, Cambridge.

IPCC (Intergovernmental Panel on Climate Change). 1993. *Climate Change 1992. The Supplementary Report to the IPCC Impacts Assessment.* Tegart, W.J.McG. & Sheldon, G.W. (Eds.), Australian Government Publishing Service, Canberra.

IPCC (Intergovernmental Panel on Climate Change) 1996. *Climate Change 1995. The Science of Climate Change. Contribution of Working Group I to the Second Assessment Report of the Intergovernmental Land on Climate Change.* Houghton, J.T., Meira Filho, L.G., Callender, B.A., Harris, A., Kattenberg, A. & Maskell, K. (Eds), Cambridge University Press, Cambridge.

Jacoby, H.D. 1990. Water quality. In: Waggoner, P.E. (Ed.) *Climate Change and US Water Resources.* Wiley, New York. 307-328.

James, I.C., Bower, B.T. & Matalas, N.C. 1969. Relative importance of variables in water resource planning. *Wat. Resour. Res.* **5**, 1165-1173.

Jenkins, A., Ferrier, R.C. & Kirby, C. (Eds) 1995. *Ecosystem Manipulation Experiments: Scientific Approaches, Experimental Design and Relevant Results.* Ecosystems Research Report No. 20. European Commission.

Jenkins, A., McCartney, M.P. & Sefton, C. 1993. *Impacts of climate change on river water quality in the United Kingdom.* Institute of Hydrology, Wallingford. Report to Department of the Environment. 39pp.

Jones, P.D. 1995. Land surface temperatures — is the network good enough? *Climatic Change* **31**, 545-558.

Jones, P.D., Briffa, K.R. & Pilcher, J.R. 1984. River flow reconstruction from tree rings in southern Britain. *J. Climatol.* **4**, 461-472.

Jones, R.G., Murphy, R.G. & Noguer, M. 1995. Simulation of climate change over Europe using a nested regional climate model. 1: assessment of control climate, including sensitivity to location of lateral boundaries. *Quart. J. Roy. Met. Soc.* **121**, 1413-1449.

Kahya, E. & Dracup, J.A. 1994. The influences of Type 1 El Niño and La Niña events on streamflows in the Pacific Southwest of the United States. *J. Climate* **7**, 965-976.

Karl, T.R., Wang, W.-C., Schlesinger, M.E., Knight, R.W. & Portman, D. 1990. A method of relating general circulation model simulated climate to the observed local climate. Part 1: seasonal statistics. *J. Climate* **3**, 1053-1079.

Karl, T.R., Bretherton, F., Easterling, W., Miller, C. & Trenberth, K. 1995. Long-term monitoring by the Global Climate Observing System (GCOS). An editorial. *Climatic Change* **31**, 135-147.

Kates, R.W. 1985. The interaction of climate and society. In: Kates, R.W., Ausubel, J.H. & Berberian, M. (Eds.), *Climate Impact Assessment: Studies of the Interaction of Climate and Society.* SCOPE 27. John Wiley, Chichester, 3-36.

Kelly, F.J. & Kelly, H.M. 1986. *What They Really Teach You at the Harvard Business School.* Piatkus, London.

Kim, J.-W., Chang, J.-T., Baker, N.L., Wilks, D.S. & Gates, W.L. 1984. The statistical problem of climate inversion: determination of the relationship between local and large-scale climate. *Mon. Weath. Rev.* **112**, 2069-2077.

Kimball, B.A. & Idso, S.B. 1983. Increasing atmospheric CO_2: effects on crop yield, water use and climate. *Agric. Wat. Manage.* **7**, 55-72.

Kimball, B.A., Mauney, J.R., Nakayama, F.S. & Idso, S.B. 1993. Effects of increasing atmospheric CO_2 on vegetation. *Vegetatio* **104/105**, 65-75.

Kirshen, P.H. & Fennessey, N.M. 1993. *Potential Impact of Climate Change upon the Water Supply of the Boston Metropolitan Area*. Report to US Environmental Protection Agency 68-W2-0018.

Kirshen, P.H. & Fennessey, N.M. 1995. Possible climate-change impacts on water supply of metropolitan Boston. *J. Wat. Resour. Plan. Manage.* **121**, 61-70.

Kite, G.W. 1989. Use of time series analysis to detect climatic change. *J. Hydrol.* **111**, 259-279.

Kite, G.W. 1993a. Analysing hydrometeorological time series for evidence of climatic change. *Nordic Hydrol.* **24**, 135-150.

Kite, G.W. 1993b. Application of a land class hydrological model to climatic change. *Wat. Resour. Res.* **29**, 2377-2384.

Klaus, J., Pflügner, W., Schmidtke, R., Wind, H. & Green, C. 1994 Models for flood hazard assessment and management. In: Penning-Rowsell, E.C. & Fordham, M. (Eds) *Floods Across Europe*. Middlesex University Press, London. 69-106.

Klemes, V. 1986. Operational testing of hydrological simulation models. *Hydrol. Sci. J.* **31**, 13-24.

Knox, J.C. 1993. Large increases in flood magnitude in response to modest changes in climate. *Nature* **361**, 430-432.

Knox, J.C. 1995. Fluvial systems since 20 000 years BP. In: Gregory, K.J., Starkel, L. & Baker, V.R. (Eds),*Global Continental Palæohydrology*. John Wiley, Chichester. 87-108.

Krasovskaia, I. & Gottschalk, L. 1992. Stability of river flow regimes. *Nordic Hydrol.* **23**, 137-154.

Kwadijk, J. 1991. Sensitivity of the River Rhine discharge to environmental change: a first tentative assessment. *Earth Surf. Processes & Landforms* **16**, 627-637.

Kwadijk, J. 1993. *The Impact of Climate Change on the Discharge of the River Rhine*. Utrecht: Faculteit Ruimtelijke Wetenschappen, Universitet Utrecht. Netherlands Geographical Studies 171. 202pp.

Kwadijk, J. & Middelkoop, H. 1994. Estimation of impact of climate change on the peak discharge probability of the River Rhine. *Climatic Change* **27**, 199-224.

Kwadijk, J. & Rotmans, J. 1995. The impact of climate change on the River Rhine: a scenario study. *Climatic Change* **30**, 397-425.

Lamb, H.H. 1982. *Climate, History and the Modern World*. Methuen, London.

Langbein, W.B. 1949. Annual runoff in the United States. *US Geol. Survey Circ.* **52**. US Dept of the Interior, Washington D.C.

Laval, K. & Polcher, J. 1993. Sensitivity of climate simulations to hydrological processes with the LMD GCM. In:*Macroscale Modelling of the Hydrosphere*. Int. Assn. Hydrol. Sci. Publ. No. 214, 133-136.

Leavesley, G.H. 1994. Modelling the effects of climate change on water resources: a review. *Climatic Change* **28**, 159-177.

Lettenmaier, D.P. & Gan, T.Y. 1990. Hydrologic sensitivities of the Sacramento-San Joaquin river basin, California, to global warming. *Wat. Resour. Res.* **26**, 69-86.

Lettenmaier, D.P., Wood, E.F. & Wallis, J.R. 1994. Hydro-climatological trends in the continental United States 1948-1988. *J. Climate* **7**, 586-607.

Lins, H. & Michaels, P.J. 1994. Increased US streamflow linked to greenhouse forcing. *Eos* **75**, 281-285.

Lockwood, J.G. 1993. Impact of global warming on evapotranspiration *Weather* **48**, 291-299.

Loomis, J. & Lee, S. 1993. Net economic value of recreational fishing on the Sacramento River in 1980. In: Dracup, J.A., Pelmuder, S.D., Howitt, R. *et al.* (Eds.), *Integrated modelling of Drought and Global Warming: Impacts on Selected California Resources*. National Institute for Global Environmental Change, University of California, Davis. Report to US Environmental Protection Agency. 97-106.

Magnuson, J.J., Crowder, L.B. & Medvick, P.A. 1979. Temperature as an ecological resource. *Amer. Zool.* **19**, 331-343.

Malanson, G.P. 1993. Comment on modelling ecological response to climatic change. *Climatic Change* **23**, 95-109.

Manabe, S. & Wetherald, R. 1987. Large-scale changes of soil wetness induced by an increase in atmospheric carbon dioxide. *J. Atmos. Sci.* **44**, 1211-1236.

Manley, G. 1974. Central England temperatures: monthly means 1659 to 1973. *Quart. J. Roy. Met. Soc.* **100**, 389-405.

Marchand, D., Sanderson, M., Howe, D. & Alpaugh, C. 1988. Climatic change and Great Lakes levels: the impact on shipping. *Climatic Change* **12**, 107-133.

Marengo, J.A. 1995. Variations and change in South American streamflow. *Climatic Change* **31**, 99-117.

Marsh, T.J. 1996. River flow and groundwater level records — the instrumented era. In: Branson, J. (Ed.), *Palæohydrology*. British Hydrological Society Occasional paper No. 7. *in press*.

Marsh, T.J., Monkhouse, R.A., Arnell, N.W., Lees, M.L. & Reynard, N.S. 1994. *The 1988-92 Drought*. Institute of Hydrology and British Geological Survey. Wallingford, Oxon. 80pp.

Martin, P., Rosenberg, N.J. & McKenny, M.S. 1989. Sensitivity of evapotranspiration in a wheat field, a forest, and a grassland to changes in climate and direct effects of carbon dioxide. *Climatic Change* **14**, 117-151.

Matyasovszky, I., Bogardi, I., Bardossy, A. & Duckstein, L. 1993. Space-time precipitation reflecting climate change. *Hydrol. Sci. J.* **38**, 539-558.

McCabe, G.J. & Wolock, D.M. 1991a. Detectability of the effects of a hypothetical temperature increase on the Thornthwaite moisture index. *J. Hydrol.* **125**, 25-35.

McCabe, G.J. & Wolock, D.M. 1991b. Effects of climatic change and climatic variability on the Thornthwaite moisture index in the Delaware River basin. *Climatic Change* **20**, 143-153.

McCaffrey, S.C. 1993. Water, politics and international law. In: Gleick, P.H. (Ed.), *Water in Crisis*. Oxford University Press, New York. 92-104.

McMahon, T.A. & Mein, R.G. 1986. *River and Reservoir Yield*. Water Resources Publications, Littleton, Colorado.

McGregor, J.L. & Walsh, K. 1994. Climate change simulations of Tasmanian precipitation using multiple nesting. *J. Geophys. Res.* **99**, D10, 20889-20905.

Mearns, L.O., Giorgi, F., McDaniel, L. & Shields, C. 1994. Analysis of daily variability of precipitation in a nested regional climate model: comparison with observations and doubled CO_2 results. *Global and Planetary Change* **10**, 55-78.

Mearns, L.O. & Rosenzweig, C. 1994. Use of nested regional climate model output with changed daily climate variability to test related sensitivity of dynamic crop models. *Preprints, Fifth Conference on Climate Variations*. American Meteorological Society, Boston. 142-155.

Mechoso, C.R. & Iribarren, G.P. 1992. Streamflow in southeastern South America and the Southern Oscillation. *J. Climate* **5**, 1535-1539.

Meyer, J.L. & Pulliam, W.M. 1992. Modification of terrestrial-aquatic interactions by a changing climate. In: Firth, P. & Fisher, S.G. (Eds.), *Global Climate Change and Freshwater Ecosystems*. Springer, Berlin. 177-191.

Miller, B.A. & Brock, W.G. 1989. *Potential impact of climate change on the Tennessee Valley Authority reservoir system*. In: Smith, J.B. & Tirpak, D.A. (Eds.), The Potential Effects of Global Climate Change on the United States. US Environmental Protection Agency, Washington D.C. Appendix A — Water Resources. 9.1-9.46.

Miller, B.A., Alavian, V., Bender, M.D. *et al.* 1993. Impacts of changes in air and water temperature on thermal power generation. In: Herrmann, R. (Ed.),*Managing Water Resources During Global Change*. Proc. Int. Symp. American Water Resources Association, Bethesda, Maryland. 439-448.

Miller, J.R. & Russell, G.L. 1992. The impact of global warming on river runoff. *J. Geophys. Res.* **97**, D3, 2757-2764.

Miller, K.A. 1989. Hydropower, water institutions and climate change. A Snake River case study. *Wat. Res. Devel.* **5**, 72-83.

Mimikou, M. & Kouvopoulos, Y.S. 1991. Regional climate change impacts: 1. Impacts on water resources. *Hydrol. Sci. J.* **36**, 247-258.

Mimikou, M., Hadjisavva, P.S., Kouvopoulos, Y.S. & Afrateos, H. 1991. Regional climate change impacts: II. Impacts on water management works. *Hydrol. Sci. J.* **36**, 259-270.

Minns, C.K. & Moore, J.E. 1993. Predicting the impact of climate change on the spatial pattern of freshwater fish yield capability in eastern Canadian lakes. *Climatic Change* **22**, 327-346.

Mitchell, J.F.B. 1990. Greenhouse warming: is the mid-Holocene a good analogue? *J. Climate* **3**, 1177-1192.

Mitchell, J.F.B. & Warrilow, D.A. 1987. Summer dryness in northern mid-latitudes due to increased CO_2. *Nature* **330**, 238-240.

Mitchell, J.F.B., Johns, T.C., Gregory, J.M. & Tett, S.F.B. 1995a. Transient climate response to increasing sulphate aerosols and greenhouse gases. *Nature* **376**, 501-504.

Mitchell, J.F.B., Davis, R.A., Ingram, W.J. & Senior, C.A. 1995b. On surface temperature, greenhouse gases, and aerosols: models and observations. *J. Climate* **8**, 2364-2386.

Mitosek, H.T. 1995. Climate variability and change within the discharge time series: a statistical approach. *Climatic Change* **29**, 101-116.

Monteith, J.L. 1965. Evaporation and environment. *Symp. Soc. Experiment. Biol.* **19**, 205-234

Moore, M.V., Folt, C.I. & Stemberger, R.S. 1995. Consequences of elevated temperatures for zooplankton assemblages in temperate lakes. *Archiv. Hydrobiol.* **268**, 1-31.

Moore, R.J. 1985. The probability-distributed principle and runoff production at point and basin scales. *Hydrol. Sci. J.* **30**, 263-297.

Murphy, J.M. 1995. Transient response of the Hadley Centre coupled ocean-atmosphere model to increasing carbon dioxide. Part I: control climate and flux correction. *J. Climate* **8**, 36-56.

Murphy, J.M. & Mitchell, J.F.B. 1995. Transient response of the Hadley Centre coupled ocean-atmosphere model to increasing carbon dioxide. Part II: Spatial and temporal structure of response. *J. Climate* **8**, 57-80.

Nash, J.E. & Sutcliffe, J.V. 1970. River flow forecasting through conceptual models. Part 1: a discussion of principles. *J. Hydrol.* **10**, 282-290.

Nash, L. 1993. Water quality and health. In: Gleick, P.H. (Ed.), *Water in Crisis*. Oxford University Press, New York. 25-39.

Nash, L. & Gleick, P.H. 1991. Sensitivity of streamflow in the Colorado Basin to climatic changes. *J. Hydrol.* **125**, 221-241.

Nash, L. & Gleick, P.H. 1993. *The Colorado River Basin and Climatic Change. The Sensitivity of Streamflow and Water Supply to Variations in Temperature and Precipitation*. Pacific Institute for Studies in Development, Environment and Security, California. Report to US Environmental Protection Agency, EPA 230-R-93-009.

National Research Council. 1977. *Climate, Climate Change and Water Supply*. US National Academy of Science, Washington D.C.

National Rivers Authority 1994. *Water: Nature's Precious Resource*. NRA, Bristol. 94pp.

Nemec, J. & Schaake, J.C. 1982. Sensitivity of water resource systems to climate variations. *Hydrol. Sci. J.* **27**, 327-343.

NERC 1975. *Flood Studies Report*. Natural Environment Research Council. HMSO, London. 5 vols.

Newson, M.D. & Lewin, J. 1991. Climate change, river flow extremes and fluvial erosion — scenarios for England and Wales. *Prog. Phys. Geogr.* **15**, 1-17.

Palutikof, J.P. 1987. Some possible impacts of greenhouse gas induced climatic change on water resources in England and Wales. In: *The Influence of Climate Change and Climatic Variability on the Hydrologic Regime and Water Resources*. Int. Assn. Hydrol. Sci. Publ. **168**, 585-596.

Palutikof, J.P., Wigley, T.M.L. & Lough, J.M. 1984. *Seasonal Climate Scenarios for Europe and North America in a High CO_2 Warmer World*. DOE/EV/10098-5. US Department of Energy, Washington D.C.

Palutikof, J.P., Gooddess, C.M. & Guo, X. 1994. Climate change, potential evapotranspiration and moisture availability in the Mediterranean basin. *Int. J. Climatol.* **14**, 853-869.

Parr, T. & Eatherall, A. 1994. *Demonstrating Climate Change Impacts in the UK: the DoE Core Model Programme*. Department of the Environment, Global Atmosphere Division, 18pp.

Parry, M.L., Carter, T.L. & Konijn, M. (Eds.) 1988. *The Impact of Climatic Variations on Agriculture*. Vol. 1, Cool and Temperate Regions. Kluwer, Dordrecht.

Penman, H.L. 1948. Natural evapotranspiration from open water, bare soil and grass. *Proc. R. Soc. London* **A193**, 120-145.

Penning-Rowsell, E.C. & Chatterton, J.B. 1977. *The Benefits of Flood Alleviation*. Gower, Aldershot.

Peterson, D. & Keller, A.A. 1990. Irrigation. In: Waggoner, P.E. (Ed.), *Climate Change and US Water Resources*. Wiley, New York. 269-306.

Pitman, A.J. 1993. The sensitivity of the land surface to the parameter values: a reassessment. In: *Macroscale Modelling of the Hydrosphere*. Int. Assoc. Hydrol. Sci. Publ. **214**, 125-132.

Poff, N.L. 1992. Regional hydrologic response to climate change: an ecological perspective. In: Firth, P. & Fisher, S.G. (Eds.), *Global Climate Change and Freshwater Ecosystems*. Springer, Berlin. 88-115.

Poff, N.L. & Allan, J.D. 1995. Functional organisation of stream fish assemblages in relation to hydrologic variability. *Ecology* **76**, 606-627.

Poff, N.L. & Ward, J.V. 1989. Implications of streamflow variability and predictability for lotic community structure: a regional analysis of streamflow patterns. *Can. J. Fish. Aquat. Sci.* **46**, 1805-1818.

Poiani, K.A. & Johnson, W.C. 1991. Global warming and prairie wetlands. *Bioscience* **41**, 611-618.

Postel, S. 1993. Water and agriculture. In: Gleick, P.H. (Ed.), *Water in Crisis*. Oxford University Press, New York. 56-66.

Press, W.H., Flannery, B.P., Teukolsky, S.A. & Vetterling, W.T. 1986. *Numerical Recipes.* Cambridge University Press: Cambridge.

Priestley, C.H.B. & Taylor, R.J. 1972. On the assessment of surface heat flux and evaporation using large-scale parameters. *Mon. Weath. Rev.* **100**, 81-92.

Racsko, P., Szeidl, L. & Semonov, M. 1991. A serial approach to local stochastic weather models. *Ecol. Modelling* **57**, 27-41.

Revelle, R.R. & Waggoner, P.E. 1983. Effects of a carbon dioxide induced climate change on water supplies in the western United States. In: *Changing Climate.* Report of the Carbon Dioxide Assessment Committee. National Academy Press, Washington D.C.

Reynard, N.S. 1993. Sensitivity of Penman-Monteith potential evaporation to climate change. In: Becker, A., Sevruk, B. & Lapin, M. (Eds) *Evaporation, Water Balance and Deposition.* Proc. Symp. Precipitation and Evaporation. Slovak Hydrometeorol. Institute, Bratislava. Vol. 3, 106-113.

Rhodes, S.L., Miller, K.A. & MacDonnell, L.J. 1992. Institutional response to climate change: water provider organizations in the Denver Metropolitan Region. *Wat. Resour. Res.* **28**, 11-18.

Riebsame, W.E. 1988. Adjusting water resources management to climate change. *Climatic Change* **13**, 69-97.

Riebsame, W.E., Meyer, W.B. & Turner, B.L. 1994. Modelling land use and cover as part of global environmental change. *Climatic Change* **28**, 45-64.

Rind, D., Goldberg, R., Hansen, J., Rosenzweig, C. & Ruedy, R. 1990. Potential evapotranspiration and the likelihood of future drought. *J. Geophys. Res.* **95**, 9983-10004.

Robock, A., Turco, R.P., Harwell, M.A. *et al.* 1993. Use of general circulation model output in the creation of climate change scenarios for impact analysis. *Climatic Change* **23**, 293-335.

Rogers, P. 1994. Assessing the socioeconomic consequences of climate change on water resources. *Climatic Change* 28, 179-208.

Rogers, P. & Fiering, M. 1990. From flow to storage. In: Waggoner, P.E. (Ed.), *Climate Change and US Water Resources.* Wiley, New York. 207-221.

Ropelewski, C.F. 1995. Long-term observations of land surface characteristics. *Climatic Change* **31**, 415-425.

Rosenberg, N.J., Kimball, B.A., Martin, P. & Cooper, C.F. 1990. From climate and CO_2 enrichment to evapotranspiration. In: Waggoner, P.E. (Ed.), *Climate Change and US Water Resources.* Wiley, New York. 151-175.

Rossow, W.B. & Cairns, B. 1995. Monitoring changes of clouds. *Climatic Change* **31**, 305-347.

Rotmans, J. 1990. IMAGE. An Integrated Model to Assess the Greenhouse Effect. Kluwer, Dordrecht.

Sælthun, N.-R., Bogen, J., Flood, M.H. *et al.* 1990. *Climate Change Impact on Norwegian Water Resources.* Norwegian Water Resources and Energy Administration. Publication 42. 34pp + app.

Sargent, R.J. & Ledger, D.C. 1992. Derivation of a 130 year runoff record from sluice records for the Loch Leven catchment, southeast Scotland. *Proc. Instn Civ. Engrs, Marine & Energy*, **96**, 71-80.

Schaake, J.C. 1990. From climate to flow. In: Waggoner, P.E. (Ed.), *Climate Change and US Water Resources.* Wiley, New York. 177-206.

Schädler, B. 1987. Long water balance time series of four important rivers in Europe — indicators for climatic changes. In: *The Influence of Climatic Change and Climatic Variability on the Hydrologic Regime and Water Resources.* Int. Assoc. Hydrol. Sci. Publ. **168**, 209-219.

Schwartz, H.E. & Dillard, L.A. 1993. Urban water. In: Waggoner, P.E. (Ed.), *Climate Change and US Water Resources.* Wiley, New York. 341-366.

Scottish Fisheries Research Station. 1992. *Description of NRA Tracking Studies*. National Rivers Authority R&D Note 33.

Sedell, J.R., Hauer, F.R., Hawkins, C.P., Reeves, G.H. & Stanford, J.A. 1990. The role of refugia in recovery from disturbances: modern fragmented and disconnected river systems. *Environ. Manage.* **14**, 711-724.

Sellers, P.J. & Hall, F.G. 1992. FIFE in 1992: results, scientific gains and future research directions. *J. Geophys. Res.* **97**, *19091-19109*

Shuter, B.J. & Post, J.R. 1990. Climate, population viability and the zoogeography of temperate fishes. *Trans. Am. Fish. Soc.* **119**, 316-336.

Shuttleworth, W.J. 1993. Evaporation. In: Maidment, D.R. (Ed.), *Handbook of Hydrology*. McGraw Hill, New York. Chapter 4.

Smith, D.I. 1993. Greenhouse climatic change and flood damages, the implications. *Climatic Change* **25**, 319-333.

Smith, J.B. & Tirpak, D. 1989. *The Potential Effect of Global Climate Change on the United States*. US Environmental Protection Agency, Washington D.C. 414pp.

Smith, K. 1990. Tourism and climate change. *Land Use Policy* **7**, 176-180.

Spence, T. & Townshend, J. 1995. The Global Climate Observing System (GCOS). An editorial. *Climatic Change* **31**, 131-134.

Stakhiv, E.Z. 1993. Water resources planning and management under climate uncertainty. In: Ballentine, T.M. & Stakhiv, E.Z. (Eds.), *Proc. First Nat. Conf. on Climate Change and Water Resources Management*. US Army Corps of Engineers. IV20-IV35.

Stamm, J.F., Wood, E.F. & Lettenmaier, D.P. 1994. Sensitivity of a GCM simulation of global climate to the representation of land surface hydrology. *J. Climate* **7**, 1218-1239.

Starkel, L. 1995. Palæohydrology of the temperate zone. In: Gregory, K.J., Starkel, L. & Baker, V.R. (Eds.), *Global Continental Palæohydrology*. Wiley, Chichester. 233-257.

Stefan, H.G. & Fang, X. 1994. Model simulations of dissolved oxygen characteristics of Minnesota lakes: past and future. *Environ. Manage.* **18**, 73-92.

Stefan, H.G. & Sinokrot, B.A. 1993. Projected global climate change impact on water temperature in five north central US streams. *Climatic Change* **24**, 353-381.

Stein, J.F. 1993. The impact of climatic change on interstate apportionment. In: Herrmann, R. (Ed.), *Managing Water Resources During Global Change*. Proc. Int. Symp. American Water Resources Association, Bethesda MD. 619-626.

Street-Perrott, F.A. & Roberts, N. 1994. Past climates and future greenhouse warming. In: Roberts, N. (Ed.), *The Changing Global Environment*. Blackwell, Oxford. 47-68.

Sweeney, B.W., Jackson, J.K., Newbold, J.D. & Funk, D.H. (1992) Climate change and the life histories and biogeography of aquatic insect in eastern North America. In: Firth, P. & Fisher, S.G. (Eds.), *Global Climate Change and Freshwater Ecosystems*. Springer, Berlin. 143-176.

Thomas, C. & Howlett, D. 1993. *Resource Politics: Freshwater and Regional Relations*. Open University Press, Buckingham.

Thompson, N., Barrie, I.A. & Ayles, M. 1981. *The Meteorological Office rainfall and evaporation calculation system (MORECS)*. Hydrological Memorandum 45, Met. O. 8, Meteorological Office, Bracknell, Berks.

Thornthwaite, C.W. 1948. An approach towards a rational classification of climate. *Geogr. Rev.* **38**, 55-94.

Tickell, C. 1993. Global warming and its effects. In: White, R. (Ed.), *Engineering for Climatic Change*. Thomas Telford, London. 1-8.

Tyree, M.T. & Alexander, J.D. 1993. Plant water relations and the effects of elevated CO_2: a review and suggestions for future research. *Vegetatio* **104/105**, 47-62.

Vaccaro, J.J. 1992. Sensitivity of groundwater recharge estimates to climate variability and change, Columbia Plateau, Washington. *J. Geophys. Res.* **97**, D3, 2821-2833.

Vehvilainen, B. & Lohvansuu, J. 1991. The effects of climate change on discharges and snow cover in Finland. *Hydrol. Sci. J.* **36**, 109-121.

Viner, D. & Hulme, M. 1993. *Construction of Climate Change Scenarios by Linking GCM and STUGE Output.* Climate Impacts LINK Technical Note 2. Climatic Research Unit. University of East Anglia, Norwich, England.

Vorosmarty, C.J., Gutowski, W.J., Person, M., Chen, T.-C. & Case, D. 1993. Linked atmosphere-hydrology models at the macroscale. In: *Macroscale Modelling of the Hydrosphere.* Int. Assoc. Hydrol. Sci. Publ. **214**, 3-27.

Walsh, J.E. 1995. Long-term observations for monitoring of the cryosphere. *Climatic Change* **31**, 369-394.

Ward, F.A. 1987. Economics of water allocation to instream uses in a fully appropriated river basin: evidence from a New Mexico wild river. *Wat. Resour. Res.* **23**, 381-392.

Water Services Association. 1994. *Waterfacts 1994.* Water Services Association, London. 62pp.

Weatherhead, E.K., Place, A.J., Morris, J. & Burton, M. 1993. *Demand for Irrigation Water.* NRA R&D Report 14. National Rivers Authority, Bristol.

Webb, B.W. 1992. *Climate Change and the Thermal Regime of Rivers.* University of Exeter, Department of Geography. Report to the Department of the Environment. 79pp.

Whetton, P.H., Fowler, A.M., Haylock, M.R. & Pittock, B. 1993. Implications of climate change due to the enhanced greenhouse gas on floods and droughts in Australia. *Climatic Change* **25**, 289-317.

Wigley, T.M.L. & Jones, P.D. 1985. Influence of precipitation changes and direct CO_2 effects on streamflow. *Nature* **314**, 149-152.

Wigley, T.M.L. & Raper, S.C.B. 1992. Implications for climate and sea level of revised IPCC emissions scenarios. *Nature* **357**, 293-300.

Wigley, T.M.L., Lough, J.M. & Jones, P.D. 1984. Spatial patterns of precipitation in England and Wales and a revised homogenous England and Wales precipitation series. *J. Climatol.* **4**, 1-26.

Wigley, T.M.L., Jones, P.D., Briffa, K.R. & Smith, G. 1990. Obtaining sub-grid-scale information from coarse-resolution general circulation model output. *J. Geophys. Res.* **95**. D2, 1943-1953.

Wilby, R. 1995. Simulation of precipitation by weather pattern and frontal analysis. *J. Hydrol.* **173**, 91-110.

Wilby, R., Greenfield, B. & Glenny, C. 1993. A coupled synoptic-hydrological model for climate change impact assessment. *J. Hydrol.* **153**, 265-290.

Wilkinson, W.B. & Cooper, D.M. 1993. The response of idealised aquifer/river systems to climate change. *Hydrol. Sci. J.* **38**, 379-390.

Wilks, D.S. 1992. Adapting stochastic weather generation algorithms for climate change studies. *Climatic Change* **22**, 67-84.

Wilson, L.L., Lettenmaier, D.P. & Skyllingstad, E. 1992. A hierarchical stochastic model of large-scale atmospheric circulation patterns and multiple station daily precipitation. *J. Geophys. Res.* **97**, D3, 2791-2809.

Wolock, D.M. & Hornberger, G.M. 1991. Hydrological effects of changes in levels of atmospheric carbon dioxide. *J. Forecasting* **10**, 105-116.

Wolock, D.M., McCabe, G.J., Tasker, G.D. & Moss, M.E. 1993. Effects of climate change on water resources in the Delaware River basin. *Wat. Resour. Bull.* **29**, 475-486.

Wood, E.F, Lettenmaier, D.P. & Zartarian, V.G. 1992. A land-surface hydrology parameterization with subgrid variability for general circulation models. *J. Geophys. Res.* **97**, D3, 2717-2728.

Wright, C.E. 1980. A study of 75 representative basins in England and Wales. In: *The Influence of Man on the Hydrological Regime with Special Reference to Representative and Experimental Basins*. Int. Assn Hydrol. Sci. Publ. **130**, 477-483.

Yates, D. & Strzepek, K. 1994a. *Potential Evapotranspiration Methods and their Impact on the Assessment of River Basin Runoff under Climate Change*. International Institute for Applied Systems Analysis Working Paper 94-46. IIASA: Laxenburg, Austria.

Yates, D. & Strzepek, K. 1994b. *Comparison of Models for Climate Change Assessment of River Basin Runoff*. International Institute for Applied Systems Analysis Working Paper 94-45. IIASA: Laxenburg, Austria.

Young, K.C. 1994. Reconstructing streamflow time series in central Arizona using monthly precipitation and tree-ring records. *J. Climate* **7**, 361-374.

Index

Note: Page references in *italics* refer to Figures; those in **bold** refer to Tables

Index compiled by Annette J. Musker